Dynamics of Cancer

Dynamics of Cancer

Incidence, Inheritance, and Evolution

STEVEN A. FRANK

Princeton University Press
Princeton and Oxford

Library of Congress Cataloging-in-Publication Data

Frank, Steven A., 1957–
Dynamics of cancer : incidence, inheritance, and
evolution / Steven A. Frank. p. ; cm. – (Princeton
series in evolutionary biology)
Includes bibliographic references and index.
ISBN 978-0-691-13365-2 (cloth : alk. paper)
ISBN 978-0-691-13366-9 (pbk. : alk. paper)
1. Carcinogenesis. 2. Cancer–Age factors.
3. Cancer–Genetic aspects. 4. Cancer–Epidemiology.
I. Title. II. Series.
[DNLM: 1. Neoplasms–etiology. 2. Age of Onset.
3. Gene Expression Regulation, Neoplastic. 4. Genetic
Predisposition to Disease. 5. Mutagenesis. 6. Stem
Cells. QZ 202 F828d 2007]
RC268.5.F63 2007
616.99′4071–dc22   2007007855

British Library Cataloging-in-Publication Data is available

Typeset by the author with TEX
Composed in Lucida Bright

Printed on acid-free paper. ∞
press.princeton.edu
Printed in the United States of America
10 9 8 7 6 5 4 3 2 1

Dynamics of Cancer

Incidence, Inheritance,

and Evolution

STEVEN A. FRANK

Princeton University Press
Princeton and Oxford

Library of Congress Cataloging-in-Publication Data

Frank, Steven A., 1957–
Dynamics of cancer : incidence, inheritance, and
evolution / Steven A. Frank. p. ; cm. – (Princeton
series in evolutionary biology)
Includes bibliographic references and index.
ISBN 978-0-691-13365-2 (cloth : alk. paper)
ISBN 978-0-691-13366-9 (pbk. : alk. paper)
1. Carcinogenesis. 2. Cancer–Age factors.
3. Cancer–Genetic aspects. 4. Cancer–Epidemiology.
I. Title. II. Series.
[DNLM: 1. Neoplasms–etiology. 2. Age of Onset.
3. Gene Expression Regulation, Neoplastic. 4. Genetic
Predisposition to Disease. 5. Mutagenesis. 6. Stem
Cells. QZ 202 F828d 2007]
RC268.5.F63 2007
616.99′4071–dc22   2007007855

British Library Cataloging-in-Publication Data is available

Typeset by the author with TEX
Composed in Lucida Bright

Printed on acid-free paper. ∞
press.princeton.edu
Printed in the United States of America
10 9 8 7 6 5 4 3 2 1

I wish I had the voice of Homer
To sing of rectal carcinoma,
Which kills a lot more chaps, in fact,
Than were bumped off when Troy was sacked.
                              —J. B. S. Haldane

Contents

PART III: EVOLUTION

PART III: EVOLUTION

Dynamics of Cancer

1

Introduction

Through failure we understand biological design. Geneticists discover the role of a gene by studying how a mutation causes a system to fail. Neuroscientists discover mental modules for face recognition or language by observing how particular brain lesions cause cognitive failure.

Cancer is the failure of controls over cellular birth and death. Through cancer, we discover the design of cellular controls that protect against tumors and the architecture of tissue restraints that slow the progress of disease.

Given a particular set of genes and a particular environment, one cannot say that cancer will develop at a certain age. Rather, failure happens at different rates at different ages, according to the age-specific incidence curve that defines failure.

To understand cancer means to understand the genetic and environmental factors that determine the incidence curve. To learn about cancer, we study how genetic and environmental changes shift the incidence curve toward earlier or later ages.

The study of incidence means the study of rates. How does a molecular change alter the rate at which individuals progress to cancer? How does an inherited genetic change alter the rate of progression? How does natural selection shape the design of regulatory processes that govern rates of failure?

Over fifty years ago, Armitage and Doll (1954) developed a multistage theory to analyze rates of cancer progression. That abstract theory turned on only one issue: ultimate system failure—cancer—develops through a sequence of component failures. Each component failure, such as loss of control over cellular death or abrogation of a critical DNA repair pathway, moves the system one stage along the progression to disease. Rates of component failure and the number of stages in progression determine the age-specific incidence curve. Mutations that knock out a component or increase the rate of transition between stages shift the incidence curve to earlier ages.

I will review much evidence that supports the multistage theory of cancer progression. Yet that support often remains at a rather vague level: little more than the fact that progression seems to follow through multiple stages. A divide separates multistage theory from the daily work of cancer research.

The distance between theory and ongoing research arose naturally. The theory follows from rates of component failures and age-specific incidence in populations; most cancer research focuses on the mechanistic and biochemical controls of particular components such as the cell cycle, cell death, DNA repair, or nutrient acquisition. It is not easy to tie failure of a particular pathway in cell death to an abstract notion of the rate of component failure and advancement by a stage in cancer progression.

In this book, I work toward connecting the great recent progress in molecular and cellular biology to the bigger problem: how failures in molecular and cellular components determine rates of progression and the age-specific incidence of cancer. I also consider how one can use observed shifts in age-specific incidence to analyze the importance of particular molecular and cellular aberrations. Shifts in incidence curves measure changes in failure rates; changes in failure rates provide a window onto the design of molecular and cellular control systems.

1.1 Aims

The age-specific incidence curve reflects the processes that drive disease progression, the inheritance of predisposing genetic variants, and the consequences of carcinogenic exposures. It is easy to see that these various factors must affect incidence. But it is not so obvious how these factors alter measurable, quantitative properties of age-specific incidence.

My first aim is to explore, in theory, how particular processes cause quantitative shifts in age-specific incidence. That theory provides the tools to develop the second aim: how one can use observed changes in age-specific incidence to reveal the molecular, cellular, inherited, and environmental factors that cause disease. Along the way, I will present a comprehensive summary of observed incidence patterns, and I will synthesize the intellectual history of the subject.

I did not arbitrarily choose to study patterns of age-specific incidence. Rather, as I developed my interests in cancer and other age-related diseases, I came to understand that age-specific incidence forms the nexus

through which hidden process flows to observable outcome. In this book, I address the following kinds of questions, which illustrate the link between disease processes and age-related outcomes.

Faulty DNA repair accelerates disease onset—that is easy enough to guess—but does poor repair accelerate disease a little or a lot, early in life or late in life, in some tissues but not in others?

Carcinogenic chemicals shift incidence to earlier ages: one may reasonably measure whether a particular dosage is carcinogenic by whether it causes a shift in age-specific incidence, and measure potency by the degree of shift in the age-incidence curve. Why do some carcinogens cause a greater increase in disease if applied early in life, whereas other carcinogens cause a greater increase if applied late in life? Why do many cancers accelerate rapidly with increasing time of carcinogenic exposure, but accelerate more slowly with increasing dosage of exposure? What processes of disease progression do the chemicals affect, and how do changes in those biochemical aspects of cells and tissues translate into disease progression?

Inherited mutations sometimes abrogate key processes of cell cycle control or DNA repair, leading to a strong predisposition for cancer. Why do such mutations shift incidence to earlier ages, but reduce the rate at which cancer increases (accelerates) with age?

Why do the incidences of most diseases, including cancer, accelerate more slowly later in life? What cellular, physiological, and genetic processes of disease progression inevitably cause the curves of death to flatten in old age?

Inherited mutations shift incidence to earlier ages. How do the particular changes in age-specific incidence caused by a mutation affect the frequency of that mutation in the population?

How do patterns of cell division, tissue organization, and tissue renewal via stem cells affect the accumulation of somatic mutations in cell lineages? How do the rates of cell lineage evolution affect disease progression? How do alternative types of heritable cellular changes, such as DNA methylation and histone modification, affect progression? How can one measure cell lineage evolution within individuals?

I will not answer all of these questions, but I will provide a comprehensive framework within which to study these problems.

Above all, this book is about biological reliability and biological failure. I present a full, largely novel development of reliability theory that accounts for biological properties of variability, inheritance, and multiple pathways of disease. I discuss the consequences of reliability and failure rates for evolutionary aspects of organismal design. Cancer provides an ideal subject for the study of reliability and failure, and through the quantitative study of failure curves, one gains much insight into cancer progression and the ways in which to develop further studies of cancer biology.

1.2 How to Read

Biological analysis coupled with mathematical development can produce great intellectual synergy. But for many readers, the mixed language of a biology-math marriage can seem to be a private dialect understood by only a few intimates.

Perhaps this book would have been an easier read if I had published the quantitative theory separately in journals, and only summarized the main findings here in relation to specific biological problems. But the real advance derives from the interdisciplinary synergism, diluted neither on the biological nor on the mathematical side. If fewer can immediately grasp the whole, more should be attracted to try, and with greater ultimate reward. Progress will ultimately depend on advances in biology, on advances in the conceptual understanding of reliability and failure, and on advances in the quantitative analysis and interpretation of data.

I have designed this book to make the material accessible to readers with different training and different goals. Chapters 2 and 3 provide background on cancer that should be accessible to all readers. Chapter 4 presents a novel historical analysis of the quantitative study of age-specific cancer incidence. Chapter 5 gives a gentle introduction to the quantitative theory, why such theory is needed, and how to use it. That mathematical introduction should be readable by all.

Chapters 6 and 7 develop the mathematical theory, with much original work on the fundamental properties of reliability and failure in biological systems. Each section in those two mathematical chapters includes a nontechnical introduction and conclusion, along with figures that illustrate the main concepts. Those with allergy to mathematics can glance briefly at the section introductions, and then move along quickly before the reaction grows too severe. The rest of the book applies the quantitative concepts of the mathematical chapters, but does so in a way that can be read with nearly full understanding independently of the mathematical details.

Chapters 8, 9, and 10 apply the quantitative theory to observed patterns of age-specific incidence. I first test hypotheses about how inherited, predisposing genotypes shift the age-specific incidence of cancer. I then evaluate alternative explanations for the patterns of age-specific cancer onset in response to chemical carcinogen exposure. Finally, I analyze data on the age-specific incidence of the leading causes of death, such as heart disease, cancer, cerebrovascular disease, and so on.

I then turn to various evolutionary problems. In Chapter 11, I evaluate the population processes by which inherited genetic variants accumulate and affect predisposition to cancer. Chapters 12 and 13 discuss how somatic genetic mutations arise and affect progression to disease. For somatic cell genetics, the renewal of tissues through tissue-specific adult stem cells plays a key role in defining the pattern of cell lineage history and the accumulation of somatic mutations. Chapter 14 finishes by describing empirical methods to study cell lineages and the accumulation of heritable change.

The following section provides an extended summary of each chapter. I give those summaries so that readers with particular interests can locate the appropriate chapters and sections, and quickly see where I present specific analyses and conclusions. The extended summaries also allow one to develop a customized reading strategy in order to focus on a particular set of topics or approaches. Many readers will prefer to skip the summaries for now and move directly to Chapter 2.

1.3 Chapter Summaries

Part I of the book provides background in three chapters: incidence, progression, and conceptual foundations. Each chapter can be read independently as a self-contained synthesis of a major topic.

Chapter 2 describes the age-specific incidence curve. That failure curve defines the outcome of particular genetic, cellular, and environmental processes that lead to cancer. I advocate the acceleration of cancer as the most informative measure of process: acceleration measures how fast the incidence (failure) rate changes with age. I plot the incidence and acceleration curves for 21 common cancers. I include in the Appendix detailed plots comparing incidence between the 1970s and 1990s, and comparing incidence between the USA, Sweden, England, and Japan. I also compare incidence between males and females for the major cancers.

I continue Chapter 2 with summaries of incidence of major childhood cancers and of inherited cancers. I finish with a description of how chemical carcinogens alter age-specific incidence. Taken together, this chapter provides a comprehensive introduction to the observations of cancer incidence, organized in a comparative way that facilitates analysis of the factors that determine incidence.

Chapter 3 introduces cancer progression as a sequence of failures in components that regulate cells and tissues. I review the different ways in which the concept of multistage progression has been used in cancer research. I settle on *progression* in the general sense of development through multiple stages, with emphasis on how rates of failure for individual stages together determine the observed incidence curve. I then describe multistage progression in colorectal cancer, the clearest example of distinct morphological and genetical stages in tumor development. Interestingly, colorectal cancer appears to have alternative pathways of progression through different morphological and genetic changes; the different pathways are probably governed by different rate processes.

The second part of Chapter 3 focuses on the kinds of physical changes that occur during progression. Such changes include somatic mutation, chromosomal loss and duplication, genomic rearrangements, methylation of DNA, and changes in chromatin structure. Those physical changes alter key processes, resulting, for example, in a reduced tendency for cell suicide (apoptosis), increased somatic mutation and chromosomal instability, abrogation of cell-cycle checkpoints, enhancement of cell-cycle accelerators, acquisition of blood supply into the developing tumor, secretion of proteases to digest barriers against invasion of other tissues, and neglect of normal cellular death signals during migration into a foreign tissue. I finish with a discussion of how changes accumulate over time, with special attention to the role of evolving cell lineages throughout the various stages of tumor development.

Chapter 4 analyzes the history of theories of cancer incidence. I start with the early ideas in the 1920s about multistage progression from chemical carcinogenesis experiments. I follow with the separate line of mathematical multistage theory that developed in the 1950s to explain the patterns of incidence curves. Ashley (1969a) and Knudson (1971) provided the most profound empirical test of multistage progression. They reasoned that if somatic mutation is the normal cause of progression, then individuals who inherit a mutation would have one less step to pass before cancer arises. By the mathematical theory, one less step shifts the incidence curve to earlier ages and reduces the slope (acceleration) of failure. Ashley (1969a) compared incidence in normal individuals and those who inherit a single mutation predisposing to colon cancer: he found the predicted shift in incidence to earlier ages among the predisposed individuals. Knudson (1971) found the same predicted shift between inherited and noninherited cases of retinoblastoma.

I continue Chapter 4 with various developments in the theory of multistage progression. One common argument posits that somatic mutation alone pushes progression too slowly to account for incidence; however, the actual calculations remain ambiguous. Another argument emphasizes the role of clonal expansion, in which a cell at an intermediate stage divides to produce a clonal population that shares the changes suffered by the progenitor cell. The large number of cells in a clonal population raises the target size for the next failure that moves progression to the following stage. I then discuss various consequences of cell lineage history and processes that influence the accumulation of change in lineages. I end by returning to the somatic mutation rate, and how various epigenetic changes such as DNA methylation or histone modification may augment the rate of heritable change in cell lineages.

Part II turns to the dynamics of progression and the causes of the incidence curve. I first present extensive, original developments of multistage theory. I then apply the theory to comparisons between different genotypes that predispose to cancer and to different treatments of chemical carcinogens. I also apply the quantitative theory of age-specific failure to other causes of death besides cancer; the expanded analysis provides a general theory of aging.

Chapter 5 sets the background for the quantitative analysis of incidence. Most previous theory fit specific models to the data of incidence curves. However, fitting models to the data provides almost no insight;

such fitting demonstrates only sufficient mathematical malleability to be shaped to particular observations. A good framework and properly formulated hypotheses express comparative predictions: how incidence shifts in response to changes in genetics and changes in the cellular mechanisms that control rates of progression. This book strongly emphasizes the importance of comparative hypotheses in the analysis of incidence curves and the mechanisms that protect against failure.

I continue Chapter 5 with the observations of incidence to be explained. I follow with simple formulations of theories to introduce the basic approach and to show the value of quantitative theories in the analysis of cancer. I finish with technical definitions of incidence and acceleration, the fundamental measures for rates of failure and how failure changes with age.

Chapters 6 and 7 provide full development of the quantitative theory of incidence curves. Each section begins with a summary that explains in plain language the main conceptual points and conclusions. After that introduction, I provide mathematical development and a visual presentation in graphs of the key predictions from the theory.

In Chapters 6 and 7, I include several original mathematical models of incidence. I developed each new model to evaluate the existing data on cancer incidence and to formulate appropriate hypotheses for future study. These chapters provide a comprehensive theory of age-specific failure, tailored to the problem of multistage progression in cell lineages and in tissues, and accounting for inherited and somatic genetic heterogeneity. I also relate the theory to classical models of aging given by the Gompertz and Weibull formulations. Throughout, I emphasize comparative predictions. Those comparative predictions can be used to evaluate the differences in incidence curves between genotypes or between alternative carcinogenic environments.

Chapter 8 uses the theory to evaluate shifts in incidence curves between individuals who inherit distinct predisposing genotypes. I begin by placing two classical comparisons between inherited and noninherited cancer within my quantitative framework. The studies of Ashley (1969a) on colon cancer and Knudson (1971) on retinoblastoma made the appropriate comparison within the multistage framework, demonstrating that the inherited cases were born one stage advanced relative to the noninherited cases. I show how to make such quantitative comparisons more simply and to evaluate such comparisons more rigorously, easing the way for more such quantitative comparisons in the evaluation of cancer genetics. Currently, most research compares genotypes only in a qualitative way, ignoring the essential information about rates of progression.

I continue Chapter 8 by applying my framework for comparisons between genotypes to data on incidence in laboratory populations of mice. In one particular study, the mice had different genotypes for mismatch repair of DNA lesions. I show how to set up and test a simple comparative hypothesis about the relative incidence rates of various genotypes in relation to predictions about how aberrant DNA repair affects progression. This analysis provides a guide for the quantitative study of rates of progression in laboratory experiments. I finish this chapter with a comparison of breast cancer incidence between groups that may differ in many predisposing genes, each of small effect. Such polygenic inheritance may explain much of the variation in cancer predisposition. I develop the quantitative predictions of incidence that follow from the theory, and show how to make appropriate comparative tests between groups that may have relatively high or low polygenic predisposition. The existing genetic data remain crude at present. But new genomic technologies will provide rapid increases in information about predisposing genetics. My quantitative approach sets the framework within which one can evaluate the data that will soon arrive.

Chapter 9 compares incidence between different levels of chemical carcinogen exposure. Chemical carcinogens add to genetics a second major way in which to test comparative predictions about incidence in response to perturbations in the underlying mechanisms of progression. I first discuss the observation that incidence rises more rapidly with duration of exposure to a carcinogen than with dosage. I focus on the example of smoking, in which incidence rises with about the fifth power of the number of years of smoking and about the second power of the number of cigarettes smoked. This distinction between duration and dosage, which arises in studies of other carcinogens, sets a classic puzzle in cancer research. I provide a detailed evaluation of several alternative hypotheses. Along the way, I develop new quantitative analyses to evaluate the alternatives and facilitate future tests.

The next part of Chapter 9 develops the second classic problem in chemical carcinogenesis, the pattern of incidence after the cessation of carcinogen exposure. In particular, lung cancer incidence of continuing cigarette smokers increases with approximately the fifth power of the duration of smoking, whereas incidence among those who quit remains relatively flat after the age of cessation. I provide a quantitative analysis of alternative explanations. Finally, I argue that laboratory studies can be particularly useful in the analysis of mechanisms and rates of progression if they combine alternative genotypes with varying exposure to chemical carcinogens. Genetics and carcinogens provide different ways of uncovering failure and therefore different ways of revealing mechanism. I describe a series of hypotheses and potential tests that combine genetics and carcinogens.

Chapter 10 analyzes age-specific incidence for the leading causes of death. I evaluate the incidence curves for mortality in light of the multistage theories for cancer progression. This broad context leads to a general multicomponent reliability model of age-specific disease. I propose two quantitative hypotheses from multistage theory to explain the mortality patterns. I conclude that multistage reliability models will develop into a useful tool for studies of mortality and aging.

Part III discusses evolutionary problems. Cancer progresses by the accumulation of heritable change in cell lineages: the accumulation of heritable change in lineages is evolutionary change.

Heritable variants trace their origin back to an ancestral cell. If the ancestral cell of a variant came before the most recent zygote, then the individual inherited that variant through the parental germline. The frequency of inherited variants depends on mutation, selection, and the other processes of population genetics. If the ancestral cell of a variant came within the same individual, after the zygote, then the mutation arose somatically. Somatic variants drive progression within an individual.

Chapter 11 focuses on germline variants that determine the inherited predisposition to cancer. I first review the many different kinds of inherited variation, and how each kind of variation affects incidence. Variation may, for example, be classified by its effect on a single locus, grouping together all variants that cause loss of function into a single class. Or variation may be measured at particular sites in the DNA sequence, allowing greater resolution with regard to the origin of variants, their effects, and their fluctuations in frequency. With resolution per site, one can also evaluate the interaction between variants at different sites. I then turn around the causal pathway: the phenotype of a variant—progression and incidence—influences the rate at which that variant increases or decreases within the population. The limited data appear to match expectations: variants that cause a strong shift of incidence to earlier ages occur at low frequency; variants that only sometimes lead to disease occur more frequently.

I finish Chapter 11 by addressing a central question of biomedical genetics: Does inherited disease arise mostly from few variants that occur at relatively high frequency in populations or from many variants that each occur at relatively low frequency? Inheritance of cancer provides the best opportunity for progress on this key question.

Chapter 12 focuses on somatic variants. Mitotic rate drives the origin of new variants and the relative risk of cancer in different tissues. For example, epithelial tissues often renew throughout life; about 80–90% of human cancers arise in epithelia. The shape of somatic cell lineages in renewing tissues affects how variants accumulate over time. Rare stem cells divide occasionally, each division giving rise on average to one replacement stem cell for future renewal and to one transit cell. The transit cell undergoes multiple rounds of division to produce the various short-lived, differentiated cells. Each transit lineage soon dies out; only the stem lineage remains over time to accumulate heritable variants. I review the stem-transit architecture of cell lineages in blood formation (hematopoiesis), gastrointestinal and epidermal renewal, and in sex-specific tissues such as the sperm, breast, and prostate.

I finish Chapter 12 by analyzing stem cells divisions and the origin of heritable variations. In some cases, stem cells divide asymmetrically, one daughter determined to be the replacement stem cell, and the other determined to be the progenitor of the short-lived transit lineage. New heritable variants survive only if they segregate to the daughter stem cell. Recent studies show that some stem cells segregate old DNA template strands to the daughter stem cells and newly made DNA copies to the transit lineage. Most replication errors probably arise on the new copies, so asymmetric division may segregate new mutations to the short-lived transit lineage. This strategy reduces the mutation rate in the long-lived stem lineage, a mechanism to protect against increased disease with age.

Chapter 13 analyzes different shapes of cell lineages with regard to the accumulation of heritable change and progression to cancer. In development, cell lineages expand exponentially to produce the cells that initially seed a tissue. By contrast, once the tissue has developed, each new mutation usually remains confined to the localized area of the tissue that descends directly from the mutated cell. Because mutations during development carry forward to many more cells than mutations during renewal, a significant fraction of cancer risk may be determined in the short period of development early in life. Once the tissue forms and tissue renewal begins, the particular architecture of the stem-transit lineages affects the accumulation of heritable variants. I analyze various stem-transit architectures and their consequences. Finally, I discuss how multiple stem cells sometimes coexist in a local pool to renew the local patch of tissue. The long-term competition and survival of stem cells in a local pool determine the lineal descent and survival of heritable variants.

Chapter 14 describes empirical methods to study cell lineages and the accumulation of heritable change. Ideally, one would measure heritable diversity among a population of cells and reconstruct the cell lineage (phylogenetic) history. Historical reconstruction estimates, for each variant shared by two cells, the number of cell divisions back to the common ancestral cell in which the variant originated. Current studies do not achieve such resolution, but do hint at what will soon come with advancing genomic technology. The current studies typically measure variation in a relatively rapidly changing aspect of the genome, such as DNA methylation or length changes in highly repeated DNA regions. Such studies of variation have provided insight into the lineage history of clonal succession in colorectal stem cell pools and the hierarchy of tissue renewal in hair follicles. Another study has indicated that greater diversity among lineages within a precancerous lesion correlate with a higher probability of subsequent progression to malignancy.

I finish Chapter 14 with a discussion of somatic mosaicism, in which distinct populations of cells carry different heritable variants. Mosaic patches may arise by a mutation during development or by a mutation in the adult that spreads by clonal expansion. Mosaic patches sometimes form a field with an increased risk of cancer progression, in which multiple independent tumors may develop. Advancing genomic technology will soon allow much more refined measures of genetic and epigenetic mosaicism. Those measures will provide a window onto cell lineage history with regard to the accumulation of heritable change—the ultimate explanation of somatic evolution and progression to disease.

Chapter 15 summarizes and draws conclusions.

PART I

BACKGROUND

2   Age of Cancer Incidence

Perturbations of the genetic and environmental causes of cancer shift the age-specific curves of cancer incidence. We understand cancer to the extent that we can explain those shifts in incidence curves. In this chapter, I describe the observed age-specific incidence patterns. The following chapters discuss what we can learn about process from these patterns of cancer incidence.

The first section introduces the main quantitative measures of cancer incidence at different ages. The standard measure is the incidence of a cancer at each age, plotted as the logarithm of incidence versus the logarithm of age. Many cancers show an approximately linear relation between incidence and age on log-log scales. I also plot the derivative (slope) of the incidence curves, which gives the acceleration of cancer incidence at different ages. The patterns of acceleration provide particularly good visual displays of how cancer incidence changes with age, giving clues about the underlying processes of cancer progression in different tissues.

The second section presents the incidence and acceleration plots for 21 different adulthood cancers. I compare the patterns of incidence and acceleration for 1993–1997 in the USA, England, Sweden, and Japan, and for 1973–1977 in the USA. Comparisons between locations and time periods highlight those aspects of cancer incidence that tend to be stable over space and time and those aspects that tend to vary. For example, many of the common cancers show declining acceleration with age: cancer incidence rises with age, but the rise occurs more slowly in later years.

The third section describes the different patterns of incidence in the common childhood cancers. The incidence of several childhood cancers does not accelerate or decelerate during the ages of highest incidence. Zero acceleration may be associated with a genetically susceptible group of individuals, each requiring only a single additional key event to lead to cancer. That single event may happen anytime during early life when the developing tissues divide rapidly, causing incidence to be equally likely over the vulnerable period.

The fourth section turns to incidence patterns in individuals that carry a strong genetic predisposition to cancer. Individuals carrying a mutation in the *APC* gene have colon cancer at a rate about three to four orders of magnitude higher than normal individuals, causing most of the susceptible individuals to suffer cancer by midlife. Susceptible individuals have an acceleration curve similar in shape to normal individuals, but shifted about 25 years earlier and slightly lower in average acceleration. Individuals carrying an *Rb* mutation have retinoblastoma at a rate about five orders of magnitude greater than normal individuals. This difference is consistent with the theory that two *Rb* mutations are the rate-limiting steps in transformation for this particular cancer, the susceptible individuals already having one of the necessary two steps.

The fifth section discusses how carcinogens alter the incidence of cancer at different ages. The best data on human cancers come from studies of people who quit smoking at different ages. Longer duration of smoking strongly increases the incidence of lung cancer. Interestingly, among nonsmokers, the acceleration of cancer does not change as individuals grow older, whereas among smokers, the acceleration tends to rise in midlife and then fall later in life. I also discuss incidence data from laboratory studies that apply carcinogens to animals. These studies show remarkably clear relationships between incidence and dose. Dose-response patterns provide clues about how mechanistic perturbations to carcinogenesis shift quantitative patterns of incidence.

The sixth section examines the different patterns of incidence between the two sexes. Males have slightly more cancers early in life. From approximately age 20 to 60, females have more cancers, mainly because breast cancer rises in incidence earlier than the other major adulthood cancers. After age 60, during the period of greatest cancer incidence, males have more cancers than females, male incidence rising to about twice female incidence. The excess of male cancers late in life occurs mainly because of sharp rises in male incidence for prostate, lung, and colon cancers. Male cancers accelerate more rapidly with age than do female cancers for lung, colon, bladder, melanoma, leukemia, and thyroid. Female cancers accelerate more rapidly for the pancreas, esophagus, and liver, but the results for those tissues are mixed among samples taken from different countries.

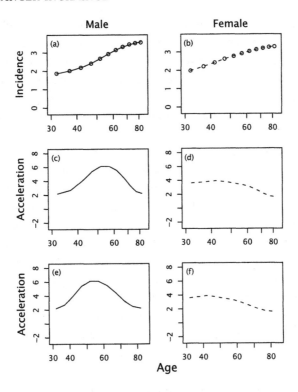

Figure 2.1 Age-specific cancer incidence and acceleration. (a,b) Age-specific incidence, the number of cancer cases for each age per 100,000 population on a log-log scale, aggregated over all types of cancer. For example, a value of 3 on the y axis corresponds to $10^3 = 1,000$ cancer cases per year, or 1 percent of the population of a given age. Circles show the data, which are tabulated in five-year intervals. I fit curves to the data with the smooth.spline function of the R statistical language, using a smoothing parameter of 0.4 (R Development Core Team 2004). (c,d) Age-specific acceleration, which is the slope (derivative) of the age-specific incidence plot at each age. I obtained the derivatives from the smoothed splines fit in the incidence plots. (e,f) The acceleration plots in the row above are transformed by changing the age axis to a linear scale to spread the ages more evenly. Data are for individuals classified racially as whites in the SEER database for USA cancer incidence, years 1973–2001 (http://seer.cancer.gov/).

2.1 Incidence and Acceleration

Age-specific incidence is the number of cancer cases per year in a particular age group divided by the number of people in that age group. Figure 2.1a,b shows age-specific incidence for USA males and females

plotted on logarithmic scales. For many types of cancer, incidence tends
to increase approximately logarithmically with age (Armitage and Doll
1954), which can be represented as $I = ct^{n-1}$, where I is incidence, t
is age, $n - 1$ is the rate of increase, and c is a constant. If we take the
logarithm of this expression, we have $\log(I) = \log(c) + (n - 1)\log(t)$.
Thus, a log-log plot of $\log(I)$ versus $\log(t)$ is a straight line with a slope
of $n - 1$.

The plots of actual cancer data rarely give perfectly straight lines on
log-log scales. The ways in which cancer incidence departs from log-
log linearity provide interesting information (Armitage and Doll 1954;
Cook et al. 1969; Moolgavkar 2004). For example, Figure 2.1a shows
the number of new cases among males per year. This is a rate, just
as the number of meters traveled per hour is a rate of motion. If we
take the slope of a rate, we get a measure of acceleration. Figure 2.1c
plots the slope taken at each point of Figure 2.1a, giving the age-specific
acceleration of cancer (Frank 2004b). If cancer accelerated at the same
pace with age, causing Figure 2.1a to be a straight line with slope $n -$
1, then acceleration would be constant over all ages, and the plot in
Figure 2.1c would be a flat line with zero slope and a value of $n - 1$ for
all ages.

Figure 2.1e takes the age-specific acceleration in Figure 2.1c and re-
scales the age axis to be linear instead of logarithmic. I do this to spread
the ages more evenly, which makes it easier to look at patterns in the
data.

The age-specific acceleration for males in Figure 2.1e shows that cancer incidence accelerates at an increasing rate up to about age 50; after 50, when most cancers occur, the acceleration declines nearly linearly. The acceleration plot for females in Figure 2.1f also shows a linear decline, starting at an earlier age and declining more slowly than
for males. The acceleration plots provide very useful complements to
the incidence plots, because changes in acceleration suggest how cancer
may be progressing within individuals at different ages (Frank 2004b).

2.2 Different Cancers

There is a vast literature on descriptive epidemiology (Adami et al.
2002; Parkin et al. 2002). Those studies examine cancer incidence at

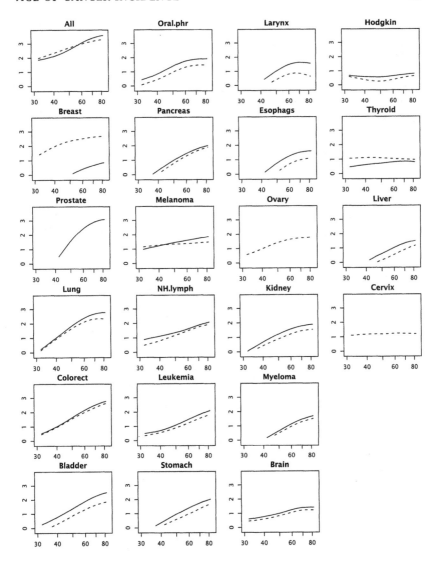

Figure 2.2   Age-specific incidence for different cancers. The curves were calculated with the same database and methods as the top row of plots in Figure 2.1.
Male cases are shown by solid lines, female cases by dashed lines. Abbreviations: *Oral.phr* for oral-pharyngeal cancer; *NH.lymph* for non-Hodgkin's lymphoma; and *Esphags* for esophageal cancer.

different times, under different environmental exposures, and in different ethnic groups. Here, I intend only to introduce the kinds of data

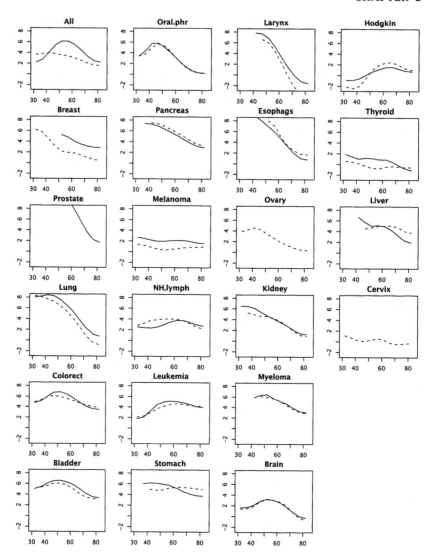

Figure 2.3 Age-specific acceleration for different cancers. The curves were calculated with the same database and methods as the bottom row of plots in Figure 2.1. Male cases are shown by solid lines, female cases by dashed lines. Abbreviations: *Oral.phr* for oral-pharyngeal cancer; *NH.lymph* for non-Hodgkin's lymphoma; and *Esphags* for esophageal cancer.

that occur, and to show some of the broad patterns that will be useful in discussing the underlying molecular and cellular processes.

Figure 2.2 plots age-specific incidence for different cancers in the USA. Solid lines show male incidences, and dashed lines show female incidences. Figure 2.3 plots the age-specific accelerations. I find it useful to look at both incidence and acceleration: incidence describes the frequency of cancer at different ages; acceleration describes how rapidly incidence changes with age at different times of life.

The acceleration plots in Figure 2.3 show nearly universal positive acceleration for these adult cancers, which means that incidence increases with age. Interestingly, the accelerations, although positive, often decline late in life (Frank 2004b). I discuss possible explanations for the late-life decline in acceleration in the following chapters.

Cancer incidence changes over time for people born in different years, perhaps because they have different lifestyles or environmental exposures (Greenlee et al. 2000). Cancer incidence also varies in different geographic locations (Parkin et al. 2002). To illustrate patterns in different times and locations, The Appendix compares incidence and acceleration of the common cancers in the USA in two time periods, 1973-1977 and 1993-1997, and in England, Sweden, and Japan in 1993-1997 (Figures A.1-A.12).

2.3 Childhood Cancers

Inherited genetic defects sometimes cause tumors in very young children (Ries et al. 1999). For example, bilateral retinoblastoma is inherited in an autosomal dominant manner (Knudson 1971). Nearly all carriers develop cancer. The early incidence and the decline in incidence with age (Figure 2.4) occur because most cell divisions in the developing retina happen in the first few years of life, and because incidence declines as the onset of disease depletes the number of susceptible but previously unaffected carriers. Unilateral retinoblastoma arises mainly in genetically normal individuals. The decline in incidence with age happens in accord with the decline in cell division in the susceptible tissue.

In testicular cancer, the early cases up to age four appear similar in pattern to the inherited early syndromes, whereas after puberty the number of cases accelerates at ages during which cell division greatly increases (Figure 2.4). Osteosarcomas increase in incidence during the ages of rapid bone elongation; these cancers decline in frequency after the teen years, with the decline in cellular division that accompanies

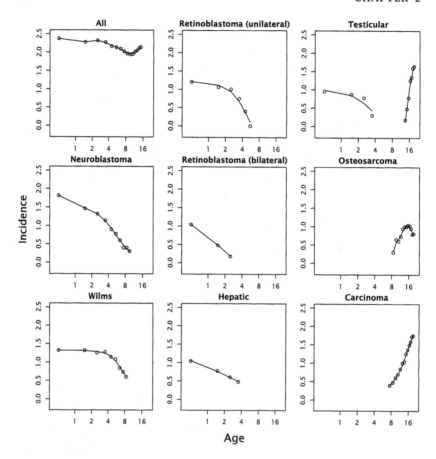

Figure 2.4   Age-specific incidence of childhood cancers on log-log scales. Incidence is given as $\log_{10}$ of the number of cases per one million population per year. Data from Ries et al. (1999) for both sexes and all races from the USA. Circles show the actual data; lines show curves fit by the smooth.spline function of R with a smoothing parameter of 0.4 (R Development Core Team 2004).

cessation of growth. Carcinomas mostly increase in incidence throughout life, because the epithelial cells continue to divide and renew those tissues at all ages.

The acceleration patterns for these cancers provide an interesting view of changes in incidence with age (Figure 2.5). The inherited syndromes have accelerations near zero or below, with a tendency to decline with age. Teen onset testicular cancer and osteosarcoma have declining

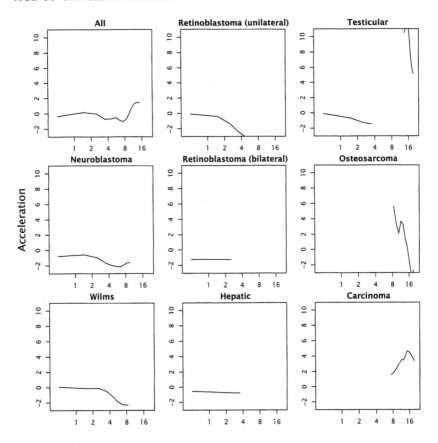

Figure 2.5   Age-specific acceleration of childhood cancers. Calculated as the slopes of the fitted splines in Figure 2.4.

accelerations, whereas carcinomas have increasing acceleration in the teen years.

2.4 Inheritance

Genetically predisposed individuals develop cancer earlier in life than do normal individuals. Ideally, we would compare age-specific incidences for different genotypes to measure how genes affect the onset of cancer.

Three problems arise in analyzing age-specific incidence curves for particular genotypes. First, currently available sample sizes tend to be small, so that we get only a rough idea of the age distribution of cases for particular kinds of genetic predisposition. Second, individuals with genetic predisposition are often identified by their cancers or the cancers of family members, causing the sample of genetically predisposed individuals to be biased and incomplete. Third, because we often do not know the base population for individuals with particular genetic tendencies, we usually cannot directly calculate incidence—the ratio of cases relative to the total number of individuals with a particular genetic predisposition over a particular time interval.

Studies vary in the extent to which they suffer from one or more of these sampling problems. Measurements will improve as better genomic techniques allow screening larger samples of individuals in an unbiased way. For now, we can look at the existing studies to get a sense of what patterns may arise.

The plots in Figures 2.4 and 2.5 use all individuals of a particular age as the base population, measuring incidence as the number of cases divided by the number of individuals in the base population. But many of those cases arose among a small subpopulation of individuals who carried particular genetic defects. It would be better to measure incidence and acceleration against the correct base population of carriers at risk for the disease. The following two examples show that, for high penetrance inherited genetic defects that lead to particular cancers, one can approximate the base population by assuming that a fixed fraction of carriers eventually develops the disease (Frank 2005).

Familial adenomatous polyposis (FAP) occurs in individuals who carry one mutated copy of the *APC* gene (Kinzler and Vogelstein 2002). This form of colon cancer can be identified during examination and distinguished from sporadic colon cancers. Figure 2.6a,b compares the incidence and acceleration for inherited and sporadic (nonfamilial) cases.

Retinoblastoma occurs as an inherited cancer in children who carry one mutated copy of the *Rb* gene (Newsham et al. 2002). Inherited cases often develop multiple tumors, usually at least one in each eye (bilateral). Retinoblastoma also occurs as a sporadic cancer, usually with only a single tumor in one eye (unilateral). Figure 2.6c,d compares the incidence and acceleration for inherited and sporadic cases.

The comparison between inherited and sporadic forms illustrates the role of genetics; the comparison between colon cancer and retinoblastoma illustrates the role of tissue development and the timing of cell division. I will return to these data in later chapters, where I consider

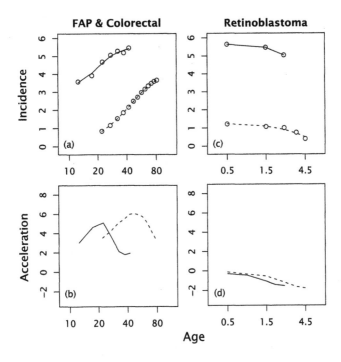

Figure 2.6   Comparison of incidence and acceleration between inherited and sporadic cancers. Incidence is given as $\log_{10}$ of the number of cases per one million population per year. Solid lines show inherited forms; dashed lines show sporadic forms. (a,b) I calculated FAP incidence by analyzing the age distribution of 129 cases combined for males and females as summarized in Ashley (1969a), from data originally presented by Veale (1965). Mutated *APC* alleles have very high penetrance for FAP, so the incidence at each age can be measured as the number of cases in an age interval divided by the fraction of individuals who had not developed the disease at earlier ages and ultimately did develop the disease. For the sporadic form, I used the incidence of colorectal cancers from the SEER database combined for white males and females from the period 1973–1977. (c,d) Inherited and sporadic forms of retinoblastoma. For the inherited form, I used 221 reported bilateral cases taken directly from the SEER database for 1973–2001. To estimate age-specific incidence, I assumed that 65 percent of carriers eventually developed bilateral tumors, based on the estimated penetrance for bilateral retinoblastoma given in Knudson (1971). The incidence in each year is approximately the fraction of cases in that year divided by the fraction of individuals in the sample who had not developed the disease in earlier years. For the sporadic form, I used the reported incidence of unilateral cases in Young et al. (1999), which is also from the SEER database. However, the SEER data do not differentiate between sporadic and hereditary unilateral cases. Based on Knudson (1971), about 75 percent of unilateral cases are sporadic cancers and about 25 percent arise from carriers who inherit a mutation. Incidence plots (a,c) from Frank (2005).

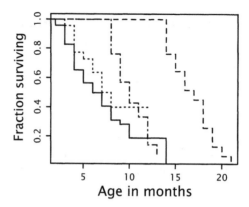

Figure 2.7 Age of lymphoma onset in mice with different mismatch repair genotypes: *Mlh3*, dashed line; *Pms2*, dot-dashed line; *Mlh1*, solid line; and *Mlh3Pms2*, dotted line. For each genotype, both alleles at each locus were knocked out. Data presented as traditional Kaplan-Meier plots, which show the fraction of mice without tumors at each age. Figure modified from Frank et al. (2005).

various hypotheses to explain these incidence and acceleration patterns. The retinoblastoma data have been particularly important in understanding how inherited and somatic mutations influence cancer progression (Knudson 1993).

Many recent laboratory studies compare the age-onset patterns of cancer between mice with different genotypes. These controlled experiments provide a clearer picture of the role of inherited genetic differences than do the uncontrolled comparisons between humans with different inherited mutations. However, most of the mouse studies have small sample sizes, making it difficult to obtain good estimates for age-onset patterns.

Figure 2.7 compares the age-onset patterns of tumors between mice with different DNA mismatch repair (MMR) genes knocked out. The figure presents Kaplan-Meier survival plots, the traditional way in which such data are reported. These plots show an association between the increase in mutation rate for defective MMR genes and a shift to earlier ages of tumor onset, in which the ordering of mutation rate is: *Mlh3* < *Pms2* < *Mlh1* ≈ *Mlh3Pms2* (Frank et al. 2005).

Analyses of laboratory experiments usually do not extract the quantitative information about age-specific incidence and acceleration from survival plots. Thus, such experiments leave unanalyzed much of the information about how particular genotypes affect the dynamics of progression. In later chapters, I show how to extract quantitative information from the traditional survival plots and use that information to test hypotheses about how genetic variants affect the dynamics of cancer progression (Frank et al. 2005).

2.5 Carcinogens

Carcinogens alter age-specific incidence patterns. The extent to which incidence patterns change depends on the dosage and the duration of exposure, and also on the age at which an individual is exposed (Druckrey 1967; Peto et al. 1991). The ways in which carcinogens change age-specific incidence may provide clues about the processes that cause cancer.

Most of the data on carcinogens come from studies of lab animals because, of course, one cannot apply carcinogens to humans in a controlled way. In later chapters, I will provide a more extensive discussion of the experimental data on carcinogens in relation to various hypotheses about the processes that lead to cancer. Here, I continue my emphasis on the patterns of incidence.

Figure 2.8 shows the best data available for carcinogen exposure in humans: the effect on lung cancer of different durations of smoking. As expected, the later the age at which individuals quit, the higher their mortality (Figure 2.8a). Interestingly, the acceleration of lung cancer is fairly constant for nonsmokers, with a slope of the log-log incidence plot for nonsmokers of about four (Figure 2.8b). For those who smoke until an age of at least 40 years, acceleration declines later in life; the late-life decline in acceleration becomes steeper with a decrease in the age at which individuals quit smoking.

Carcinogens applied to lab animals allow controlled measurement of dosage and incidence. In the largest study, Peto et al. (1991) measured the age-specific incidence of esophageal tumors in response to chronic exposure to N-nitrosodiethylamine (NDEA). Exposure of inbred rats began at about six weeks of age and continued throughout life. The data fit well to

$$I = nbt^{n-1}, \tag{2.1}$$

where I is the standard measure of age-specific incidence, b is a constant depending on dosage, t measures in years the duration of carcinogen

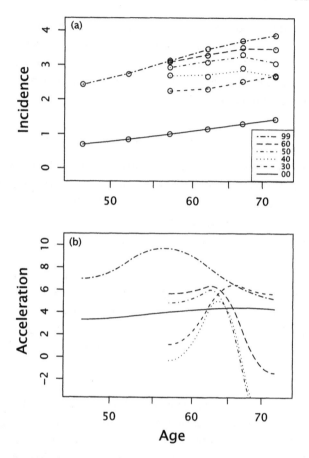

Figure 2.8 Fatal lung cancer in males for groups that quit smoking at different ages. The six curves defined in the legend show individuals who never smoked (quit at age 0), individuals who quit at ages 30, 40, 50, and 60, and individuals who never quit (shown as age 99). (a) Age-specific mortality per 100,000 population on a $\log_{10}$ scale versus age scaled logarithmically. Data extracted from Figure 2 of Cairns (2002), originally based on the analysis in Peto et al. (2000). Most cases of lung cancer are fatal, so these mortality data provide a good guide to incidence, advanced slightly in age because of the lag between the origin of the cancer and death. Curves fit to the observations (circles) by the smooth.spline function (R Development Core Team 2004), with a smoothing parameter of 0.3. (b) Age-specific acceleration calculated as the derivative (slope) of the smoothed curves fit in (a). Some of the curves in (a) are based on only four observed points, causing the fitted curves to be sensitive to the level of smoothing; the plotted accelerations in (b) for those curves should be regarded only as qualitative guides to the general trends in the data.

exposure until tumor onset, and *n* determines the scaling of incidence

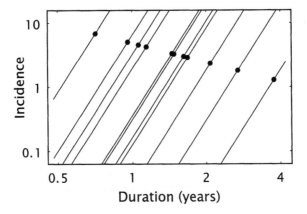

Figure 2.9   Age-specific incidence of tumor onset as a function of duration of exposure to a carcinogen. The circles show the observed median duration, the time until one-half of the experimental rats has esophageal tumors in response to chronic exposure to *N*-nitrosodiethylamine (NDEA) in drinking water (Peto et al. 1991). Each observed median corresponds to a group of rats treated with a different dosage, as shown in Figure 2.10. For each observed median, I calculated the incidence line from Eq. (2.2). These calculated lines matched well the observed age-specific incidences in each experimental group (Peto et al. 1991).

with time. Peto et al. (1991) showed mathematically that the constant b is related to m, the median duration of carcinogen exposure to tumor onset, as

$$b = -\ln(0.5)\, m^{-n} = 0.693 m^{-n}.$$

Later I will show how to derive this result. From the laboratory observations, Peto et al. (1991) estimated $n = 7$, so we can describe age-specific incidence for this experiment as

$$I = 4.85 m^{-7} t^6,$$

and, on a log-log scale,

$$\log(I) = \log(4.85) - 7\log(m) + 6\log(t). \tag{2.2}$$

This equation and Figure 2.9 show that the median, m, sets the pattern of incidence.

In the study by Peto et al. (1991), the observed relation between median duration and dosage followed the classical dose-response formula given by Druckrey (1967),

$$k = d^r m^n, \tag{2.3}$$

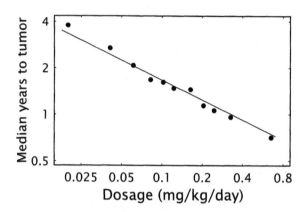

Figure 2.10   Esophageal tumor dose-response line. The circles show the same observed median durations as in Figure 2.9. Here, each median duration is matched to the dosage level for that experimental group of rats. The line shows the excellent fit to the Druckrey formula expressed in Eq. (2.4), with $r = 3$, $n = 7$, $k = 0.036$, and a slope of $-r/n = -1/s = -1/2.33$. Data from Peto et al. (1991).

where k is a constant measured in each data set; d is dosage given in this experiment as mg/kg/day; r determines the rate of increase in incidence with dosage at a fixed duration; m is the median duration; and $n - 1$ is the exponent on duration in Eq. (2.1) that fits the observed age-specific incidences. The Druckrey formula is often given as $k = dm^s$, which is equivalent to Eq. (2.3) with $s = n/r$ and a different constant value, k.

Because median time to onset captures the patterns in the data, dose-response experiments are usually summarized by plotting the medians in response to varying dosage levels. We get the expected dose-response relation by rearranging the Druckrey formula in Eq. (2.3) as

$$\log (m) = (1/n) \log (k) - (r/n) \log (d) . \qquad (2.4)$$

Figure 2.10 shows the close experimental fit to this dose-response equation obtained by Peto et al. (1991). Figure 2.11 summarizes eight earlier experiments that also showed a close fit to the Druckrey formula.

2.6 Sex Differences

Males and females have different patterns of cancer incidence. The most obvious differences occur in the reproductive tissues. For example, the breast and prostate account for a significant fraction of all cancers, as shown in Figure 2.2.

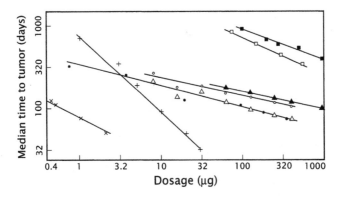

Figure 2.11 Dose-response lines from a variety of animal experiments. For each experiment, I list the slope of the line, $-r/n = -1/s$, from Eq. (2.4): ($\times$) methylcholanthrene applied to mouse skin three times per week, skin tumors with slope of $-1/2.1$; (+) 4-dimethylaminoazobenzene fed to rats in daily diet (dosage multiplied by 1000), liver tumors with slope of $-1/1.1$; (filled circle) 3,4-benzopyrene applied to mouse skin three times per week, skin tumors with slope of $-1/4.0$; (open triangle) methylcholanthrene given as a single subcutaneous injection to mice, duration measured as time after exposure, sarcomas with slope of $-1/4.0$; (open circle) 1,2,5,6-dibenzanthracene given as a single subcutaneous injection to mice, sarcomas with slope of $-1/4.7$; (filled triangle) 3,4-benzopyrene, single subcutaneous injection to mice, sarcomas with slope of $-1/4.7$; (open square) diethylnitrosamine fed to rats in daily diet, liver tumors with slope of $-1/2.3$; (filled square) dimethylaminostilbene fed to rats in daily diet, ear duct tumors with slope of $-1/3.0$. Redrawn from Figure 9 of Druckrey (1967).

Apart from the reproductive tissues, other distinctive patterns occur in the incidence of cancer in males and females. The left column of Figure 2.12 shows that, over all cancers, the relative age-specific incidences follow the same curve in different time periods and in different geographic areas. The curves show the ratio of male to female incidence rate at each age. Early in life, males have a slight excess of cancers. From roughly age 20 to 60, females have an excess of cancers, with a distinctive valley in the male:female ratio at about 40 years of age. After age 60, during which most cancers occur, males have a significant excess of cancers, rising to about twice the rate of female cancers.

Part of the aggregate pattern over all cancers can be explained by breast cancer, which occurs at a relatively high rate earlier in life than the other common cancers. The relatively high rate of breast cancer in midlife causes a female excess in the middle years, which appears

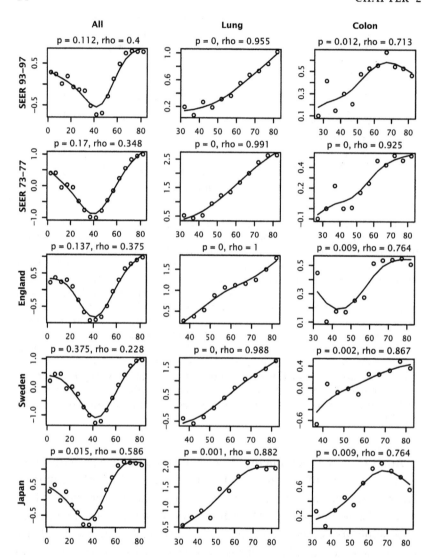

Figure 2.12   Ratio of male to female age-specific incidence. The y axis shows male incidence rate divided by female incidence rate for each age, given on a $\log_2$ scale. This scaling maps an equal male:female incidence ratio to a value of zero; each unit on the scale means a two-fold change in relative incidence, with negative values occurring when female incidence exceeds male incidence. Each plot shows the Spearman's rho correlation coefficient and p-value; a p-value of zero means $p < 0.0005$. Positive correlations occur when there is an increasing trend in the ratio of male to female incidence with increasing age. Note that the scales differ between plots, using the maximum range of the data to emphasize the shapes of the curves. The data are the same as used in Figures A.1–A.11.

as a depression in the male:female incidence ratio in the left column of Figure 2.12. Prostate and lung cancers also influence the aggregate male:female ratio—these cancers rise strongly in later years and occur only (prostate) or mostly (lung) in males.

Figures A.13–A.18 in the Appendix show the male:female ratios for the major adult cancers. The plots highlight two kinds of information. First, the values on the y axis measure the male:female ratio. Second, the trend in each plot shows the relative acceleration of male and female incidence with age. For example, in Figure 2.12, the positive trend for lung cancer shows that male incidence accelerates with age more rapidly than does female incidence, probably because males have smoked more than females, at least in the past.

Figures A.13–A.18 show that positive trends in the male:female incidence ratio also occur consistently for colon, bladder, melanoma, leukemia, and thyroid cancers. Negative trends may occur for the pancreas, esophagus, and liver, but the results for those tissues are mixed among samples taken from different countries. Simple nonlinear curves seem to explain the patterns for the stomach and Hodgkin's cancers, and maybe also for oral-pharyngeal cancers.

The patterns of relative male:female incidence probably arise from differences between males and females in exposure to carcinogens, in hormone profiles, or in patterns of tissue growth, damage, or repair. At present, the observed patterns serve mainly to guide the development of hypotheses along these lines.

2.7 Summary

This chapter summarized patterns of cancer incidence. The best theoretical framework to explain those patterns arises from the assumption that cancer progresses through multiple stages. Before turning to multistage theory and its connections to the data on incidence, it is useful to consider the observations on how cancer develops within individuals with regard to stages of progression. The next chapter summarizes observations of multistage progression.

3 Multistage Progression

Several checks prevent uncontrolled proliferation of cells. Normal cells commit suicide when they cannot pass various quality control tests; a built-in counting mechanism limits the number of times a cell can divide; structural rigidity and physical partitions in tissues prevent expansion of abnormal cellular clones. In this chapter, I describe how cancer develops by sequential changes to cells and tissues that bypass these normal checks on tissue growth.

The first section defines the word *progression* to include all of the changes that transform cells from normal to cancerous. Earlier literature split the stages of transformation into initiation, promotion, and progression to metastasis. Some tumors may develop through these particular stages, but those stages can be difficult to discern and are not universal. So I use *progression* in the general sense of development from the first to the final stages.

The second section considers the meaning of the commonly used phrase *multistage progression.* I focus on how the rates of change in progression affect the age-onset patterns of cancer. In this framework, the multiple stages of progression describe the rate-limiting steps. I use this framework in later chapters to formulate and test quantitative hypotheses about how particular events affect cancer.

The third section summarizes multistage progression in colorectal cancer. That cancer provides the clearest example of distinct morphological and genetic stages in tumor development.

The fourth section describes alternative pathways of multistage progression in colorectal cancer. The distinct morphological and genetic pathways are probably governed by different rate processes. In general, cancer of a particular tissue may be heterogeneous with regard to the pathways and rate processes of progression.

The fifth section provides a transition into the second half of the chapter, in which I summarize the kinds of changes that accumulate during progression.

The sixth section focuses on the physical changes during progression. Such changes include somatic mutation, chromosomal loss and duplication, genomic rearrangements, changes in chromatin structure and methylation of DNA, and altered gene expression.

The seventh section lists the key processes that change in progression. These changes include reduced tendency for cell suicide (apoptosis), increased somatic mutation and chromosomal instability, abrogation of cell-cycle checkpoints, enhancement of cell-cycle accelerators, acquisition of blood supply into the developing tumor, secretion of proteases to digest barriers against invasion of other tissues, and neglect of normal cellular death signals during migration into a foreign tissue.

The eighth section examines the pattern by which changes accumulate over time. The major rate-limiting changes may accumulate sequentially within a single dominant tumor cell lineage. Alternatively, different cell lineages may progress via different pathways, leading to a tumor with distinct cell lines that diverged early in progression. In distant metastases, the colonizing migrant cells may all derive from a single dominant cell lineage in a late-stage localized tumor. By contrast, metastatic migrants may emerge from different developmental stages of the primary tumor or from different cell lineages within the primary tumor, causing genetically distinct metastases. In general, cell lineage histories play a key role in understanding the nature of progression.

3.1 Terminology

Tumors develop by progression through a series of stages. Experimental studies that apply carcinogens to animals typically distinguish initiation as starting the first stages in development and promotion as stimulating the following stages (Berenblum 1941). The initiator chemicals often cause mutation; the promoter chemicals often increase cell division (Lawley 1994). The word *progression* in experimental studies is usually confined to the final developmental stages of cancer that follow promotion.

In natural tumors, there may sometimes be stages corresponding to initiation and promotion, but those stages can be highly variable, difficult to discern, and poor descriptors for particular stages in development (Iversen 1995). In all cases, *progression* nicely describes progress through a sequence of developmental stages. I use the word *progression* in this general sense of development from the first to the final stages. Within the broad sweep of progression, it may sometimes be useful to distinguish stages of initiation, promotion, and final progression to metastasis.

3.2 What Is Multistage Progression?

This question has led to confusion. Some people aim for the ordered list of necessary changes to cellular genomes and to tissues that cause aggressive cancers. Others emphasize the controversial hypothesis that two processes occur: initiation by somatic mutation as a first stage, and promotion by mitotic stimulation as a second stage.

There is no single correct way to pose the question. The listing of specific changes sets a useful although perhaps rather difficult goal. The testing of the particular two-stage hypothesis of initiation and promotion has focused on experimental studies of carcinogens in laboratory animals; the two-stage hypothesis is probably too narrow to provide a general framework for cancer development.

I focus on how the dynamics of progression within individuals affects the age-onset patterns in populations. Biochemical changes that do not affect rates of progression can be ignored in dynamical analyses, even though they may be very important for understanding physiological changes and for analyzing which drugs succeed or fail in chemotherapy.

In focusing on rate processes, I sacrifice comprehensive understanding of all aspects of cancer. In return for that sacrifice, I gain a coherent framework that gives meaning to the common but often vague assertion that some particular genetic change or biochemical event causes cancer: in the dynamical framework of multistage progression, causing cancer means shifting the age-incidence curve. With this quantitative framework, we can formulate and test hypotheses about how particular events affect cancer.

My quantitative emphasis on progression and incidence, and on testable hypotheses, means that I will not attempt to cover all aspects of progression in a comprehensive way (see Weinberg 2007). In this chapter, I give just enough background to set the stage for formulating a quantitative framework and testing simple hypotheses.

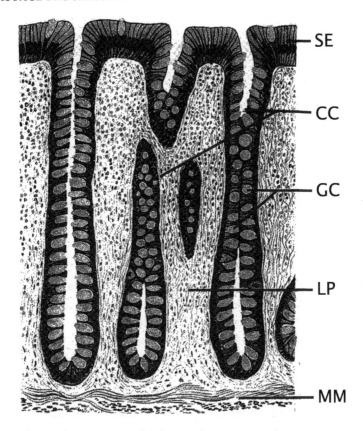

Figure 3.1   Morphology of normal colon tissue. Labels show surface epithelium (SE), colon crypts (CC), goblet cells (GC), lamina propria (LP), and muscularis mucosa (MM). The crypts open to the surface epithelium—in this cross section, some of the crypts appear partially or below the surface. From Kinzler and Vogelstein (2002), original published in Clara et al. (1974).

3.3 Multistage Progression in Colorectal Cancer

Colorectal cancer provides a good model for the study of morphological and genetic stages in cancer progression (Kinzler and Vogelstein 2002). Various precancerous morphologies can be identified, allowing tissue samples to be collected and analyzed genetically. Figure 3.1 shows the morphology of normal colon tissue. The epithelium has about 10^7 invaginations, called crypts. Cells migrate upward to the epithelial surface from the dividing stem cells and multiplying daughter cells at

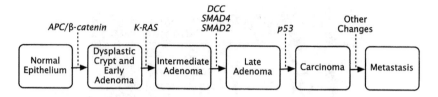

Figure 3.2 Morphology of colorectal cancer progression. This classical pathway is characterized by traditional adenoma morphology, slow progression, high adenoma:carcinoma ratio, frequent chromosomal instability and aneuploidy, and rare microsatellite instability. Particular genetic changes often associate with morphological stage, suggesting that the genetic changes play an important role in driving progression. Approximately 50–85 percent of colorectal cancers follow this pathway. Redrawn from Figure 3 of Fearon and Vogelstein (1990).

the base of the crypt. Migrating cells move from the base to the surface in about 3–6 days. Normal cells die at the surface, replaced by the continuous stream of new cells from below.

Most colorectal cancers progress through a series of morphological stages (Figure 3.2). In the first histological signs, one or more crypts show accumulation of excess cells at the surface. The cells in these aberrant crypt foci may appear normal, forming hyperplastic tissue, or the cells may have abnormal intracellular and intercellular organization, forming dysplastic tissue. As excess cells accumulate, visible polyps grow and protrude from the epithelial surface.

If the polyp is dysplastic, the tumor is called an adenoma. Adenomas tend to become more dysplastic as they grow. If the polyp is hyperplastic, it usually does not follow the classical pathway to cancer in Figure 3.2, but may occasionally follow an alternative route, as discussed later.

Early Stages

What change causes cells to accumulate at the epithelial surface and initiate adenomatous growth? Mutation of the APC regulatory pathway appears to be the first step (Kinzler and Vogelstein 2002). APC represses β-catenin, which may have two different consequences for cellular growth. First, β-catenin may enhance expression of c-Myc and other proteins that promote cellular division. Second, β-catenin may play a role in cell adhesion processes, effectively increasing the stickiness of surface epithelial cells. In either case, repression of β-catenin reduces the tendency for abnormal tissue expansion.

APC expression rises and represses β-catenin as cells migrate from the base of crypts toward the epithelial surface. Rise in APC expression and repression of β-catenin associate with increased apoptosis as cells approach the surface. Loss of surface cells is necessary to balance production from the base of crypt.

In tumors, mutations in *APC* usually include domains involved in binding β-catenin; abrogation of APC binding releases β-catenin from the suppressive effects of APC (Kinzler and Vogelstein 2002). Both *APC* alleles are probably mutated in most tumors, consistent with the hypothesis that lack of functional APC releases suppression of β-catenin and leads to adenomatous growth.

The occasional tumors that lack *APC* mutations frequently have β-catenin mutations that resist repression by APC (Jass et al. 2002b; Kinzler and Vogelstein 2002). β-catenin resistance requires that only one allele mutate to escape suppression by APC.

Disruption of the APC pathway may be sufficient to start a small adenomatous growth. Two lines of evidence point to disruption of the APC pathway as an early, perhaps initiating event in carcinogenesis (Kinzler and Vogelstein 1996, 2002). First, *APC* mutations occur as frequently in small, benign tumors as they do in cancers. By contrast, mutations in other genes commonly altered in colorectal cancers, such as *p53* and *K-RAS*, appear only later in tumor progression (Figure 3.2). Second, *APC* mutations occur in the earliest stages of aberrant crypts, consistent with the hypothesis that the first steps of stickiness and lack of cell death at the epithelial surface arise from disruption of the APC pathway.

GROWTH BEYOND SMALL ADENOMAS

Mutation of a *RAS* gene often occurs among the next genetic events of progression (Kinzler and Vogelstein 2002). Among early adenomas less than 1cm, fewer than 10 percent had mutations to either *K-RAS* or *N-RAS*, whereas more than 50 percent of adenomas greater than 1cm and carcinomas had a mutation to one of these genes. Mutations usually occur in *K-RAS* but occasionally in *H-RAS*. The *RAS* family acts oncogenically, with a mutation to a single allele sufficient to cause progression.

The strong tendency for *APC* mutations to appear in early morphological stages and *RAS* mutations to occur only in later morphological stages suggests that the order of the mutational steps plays an important role in colorectal carcinogenesis.

DEVELOPMENT OF LATE ADENOMAS

As adenomas continue to grow and begin to show great histological disorder, they tend to lose parts of 18q, the long arm of chromosome 18 (Kinzler and Vogelstein 2002). Only about 10 percent of early and intermediate adenomas have 18q chromosomal loss, whereas about 50 percent of late adenomas and 75 percent of carcinomas have 18q loss. These observations suggest one or more additional genetic events associated with continuing morphological progression through late adenoma and early carcinoma stages.

Limited evidence points to one or more of the genes *DCC, SMAD4,* and *SMAD2* in 18q21 as playing a role in carcinogenesis. DCC is a surface protein with extensive homology to other cell adhesion and surface glycoprotein molecules. Loss of *DCC* often occurs in cancers, suggesting DCC acts as a tumor suppressor. SMAD4 and SMAD2 may interact with the transforming growth factor beta (TGFβ) pathway. The TGFβ pathway often suppresses normal cellular growth, so loss of response to this pathway may release developing tumors from suppressive signals.

TRANSITION TO CANCER

Loss of functional p53 by damage to both alleles drives progression to carcinomas (Kinzler and Vogelstein 2002). p53 suppresses cell division or induces apoptosis in response to stress or damage. Cancerous growth usually requires release from p53's protective control over cellular birth and death. *p53* is on the short arm of chromosome 17, region 17p13. Allelic losses on 17p occur in less than 10 percent of early or intermediate stage adenomas, increasing to about 30 percent in late adenomas and rising to about 75 percent in cancers.

Other genetic changes probably arise during progression. During metastasis, adaptation of cancerous tissues likely occurs as the tissues become aggressive, migrate, and greatly alter the environment in which they live. Such adaptations must often depend on genetic changes.

Chromosomal Instability

About 85 percent of colorectal tumors have major chromosomal aberrations. Often, part of a chromosome or a whole chromosome is lost (Rajagopalan et al. 2003). A lost chromosome is usually replaced by duplication of the remaining chromosome from the original pair. Duplication creates two copies of the same allele at a locus, with loss of one of the original parental alleles. This is called *loss of heterozygosity*, or LOH, because the remaining duplicated pair is homozygous.

LOH accelerates the genetic changes that drive carcinogenesis (Nowak et al. 2002). For example, a mutation to one allele of *p53* leaves one original copy intact. By itself, the single mutation to one allele often does not cause severe problems. But the good copy may disappear if its chromosome is lost, and the remaining chromosome duplicates leaving two copies of the mutated allele. In chromosomally unstable genomes, chromosomal losses causing LOH happen much more rapidly than do typical mutations. The common genetic pathway of change is often a mutation to one allele at a low rate followed by loss of the other allele by LOH at a relatively rapid rate.

Chromosomal instability (CIN) arises from mutations and other genomic changes that abrogate the normal controls on chromosome duplication and segregation in mitosis (Rajagopalan et al. 2003). Because CIN increases the rate at which genetic changes occur, CIN can accelerate the sequence of genetic events that drive carcinogenesis. Most tumors of solid tissue have CIN. But it remains controversial whether CIN arises early in carcinogenesis and thus plays a key role in driving genetic change, or CIN develops late in tumorigenesis as the genome becomes increasingly disrupted by the later stages of carcinogenesis. Probably there are pathways of progression that depend on CIN and those that do not.

3.4 Alternative Pathways to Colorectal Cancer

Microsatellite Instability

Approximately 15 percent of colorectal tumors do not have CIN or widespread chromosomal abnormalities (Rajagopalan et al. 2003). Instead, these tumors usually have mutations in their mismatch repair (MMR) system, a component of DNA repair (Boland 2002). Loss of MMR

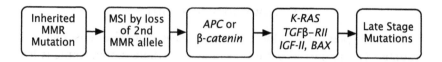

Figure 3.3   Genetic changes in HNPCC progression. Approximately 2–4 percent of colorectal cancers follow this pathway.

causes increased mutation in repeated DNA sequences, such as those in microsatellite regions. This failure to repair mismatches in repeats causes repetitive microsatellites to change their length at a much higher rate than normal during DNA replication. The observed fluctuations in microsatellite length lead to the name *microsatellite instability* (MSI) for defects in MMR. Genes with repetitive sequences seem to be at greater risk for mutation in MSI tumors.

Most colorectal tumors have either MSI or CIN, but not both. Some form of accelerated mutation may be needed for progression to aggressive colorectal cancer (Jass et al. 2002a; Kinzler and Vogelstein 2002).

HNPCC PATHWAY

Individuals who inherit defects in MMR develop hereditary nonpolyposis colorectal cancer (HNPCC) as well as other cancers that together make up Lynch's syndrome (Boland 2002). Some of the genetic steps in HNPCC progression and the rates of transition between stages differ from the classical pathway (Figure 3.3).

Typically, individuals inherit one defective allele at a locus involved in MMR. Heterozygous cells are usually normal for MMR. A somatic mutation to the second allele at the affected locus leads to loss of function in a component of the MMR system. The elevated rate of mutation causes MSI and frameshift mutations in genes with repeated sequences.

Mutation to *APC* or *β-catenin* initiates adenomatous growth. With MSI, the mutational spectrum to these genes differs from the classical pathway, which often begins with a mutated copy of *APC* followed by an LOH event to knock out both functional copies of the gene. In HNPCC, there are more mutations to *β-catenin* instead of *APC*, and mutations to *APC* more often result from frameshifts in repetitive regions caused by failure of MMR (Jass et al. 2002a, 2002b). These differences are consistent with the observation that MMR deficient tissues rarely have CIN and

LOH. In the absence of LOH, two separate mutations to *APC* are needed, whereas only one mutation to *β-catenin* is needed. This may explain why there is a rise in the ratio of *β-catenin* to *APC* initiating mutations in HNPCC.

The morphological sequence in HNPCC follows the classical pathway. In the classical pathway, the adenoma to carcinoma ratio is about 30:1. By contrast, HNPCC patients have an adenoma to carcinoma ratio of about 1:1 (Jass et al. 2002b). This suggests much faster progression from adenoma to carcinoma in HNPCC, probably driven by the high somatic mutation rate in MSI cells.

The spectrum of later mutations in HNPCC differs from later mutations in the classical pathway (Jass et al. 2002b). HNPCC tumors have less LOH. The *K-RAS* mutation frequency is about the same, but HNPCC may have fewer *p53* mutations, and more mutations in various growth-related genes with repetitive sequences, including *TGFβ-RII*, *IGF-II*, and *BAX*.

In another study, Rajagopalan et al. (2002) found that 61 percent of 330 colorectal tumors had either a *BRAF* or *K-RAS* mutation, but a tumor never had mutations in both genes. Mutually exclusive mutation of these genes supports the suggestion that they have similar effects in tumorigenesis (Storm and Rapp 1993). The ratio of *BRAF* to *K-RAS* mutations was significantly higher in MMR deficient cancers compared to MMR proficient cancers. This difference in mutation frequency again supports the idea that particular aberrations in DNA repair affect the mutation spectrum of tumors, although the functional changes caused by different mutations may sometimes be similar.

HYPERMETHYLATION

Some colorectal cancers accumulate changes in gene expression by hypermethylation of promoter regions, which can suppress transcription. Commonly hypermethylated genes in colorectal cancers include *p14, p16, hMLH1, MGMPT,* and *HPP1* (Jass et al. 2002a; Issa 2004).

Jass et al. (2002b) proposed multiple pathways to cancer via hypermethylation, accounting for up to 40 percent of all colorectal cancers. These arguments are, at present, based on limited sample sizes. But the existing data do hint at interesting hypotheses about alternative pathways.

A two-step mechanism may begin carcinogenesis in all hypermethylation pathways: reduction of apoptosis followed by increase in somatic mutation (Jass et al. 2002a). The order may be important. High somatic mutation rate in cells with normal apoptotic processes may often lead to increased cell death rather than accumulation of genetic change, because normal cells often undergo apoptosis if they cannot repair genetic damage. If apoptosis is lost first, then somatic mutations can be maintained.

With loss of apoptosis, cells accumulate in the aberrant crypt. In typical hypermethylation pathways, cellular accumulation causes hyperplastic growth with a characteristic sawtoothed or serrated morphology (Jass et al. 2002b; Jass 2003; Park et al. 2003). Hyperplasia means that the aberrant tissue retains a more or less orderly internal structure, whereas dysplasia means disordered cellular organization in the aberrant tissue.

About 95 percent of aberrant crypt foci are hyperplastic and serrated (Jass et al. 2002a). The other 5 percent are dysplastic and lack serration. The dysplastic group may often follow the classical pathway in Figure 3.2 via mutation of the APC pathway, which may abrogate apoptosis and cause accumulation of cells at the top of aberrant crypts (Kinzler and Vogelstein 2002). By contrast, hyperplastic crypts seem to accumulate cells lower down in the crypt, suggesting an alternative to *APC* mutation as an initiating event that abrogates apoptosis (Jass et al. 2002a). Alternative initiating events that interfere with apoptosis include *K-RAS* mutation and hypermethylation silencing of *HPP1/TPEF*.

Most hyperplastic aberrant crypts do not progress. However, a subsequent disruption of the DNA repair system leads to elevated somatic mutation rates, and may drive the tissue through the next stages of progression. Morphologically, serrated and hyperplastic precursor lesions sometimes show heterogeneous dysplastic outgrowths, such as serrated adenomas. Those dysplastic outgrowths usually have some form of elevated mutation, and progress relatively rapidly to cancer, causing a low adenoma to carcinoma ratio for this pathway (Jass et al. 2002a).

Jass and colleagues describe two hypermethylation syndromes. The two syndromes can be distinguished by the mechanism that elevates somatic mutation rates (Jass et al. 2002b, 2002a).

In the first syndrome, promoter methylation of *hMLH1* disrupts the MMR system, leading to high somatic mutation rate and high levels of microsatellite instability (MSI-H). These cases do not have inherited *hMLH1*

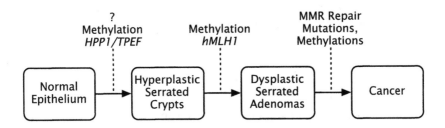

Figure 3.4 Morphological sequence in hypermethylated MSI-H cancers. Up to 15 percent of colorectal cancers follow this pathway.

mutations and differ significantly from the HNPCC pathway. Although there is much variation, the sequence in Figure 3.4 may be typical for MSI-H tumors that are not HNPCC. Many common attributes of the classical pathway are rare in this sequence. For example, these cancers have relatively low frequencies of mutations to the APC pathway, suggesting some other initiating event such as apoptotic loss via methylation of *HPP1/TPEF*. These cancers also have fewer mutations to *K-RAS* and *p53*, and usually do not have chromosomal instability or significantly altered karyotypes.

The second hypermethylation syndrome follows the same morphological pathway in Figure 3.4, but has little or no MSI. The early hyperplastic, serrated morphology suggests an initiating event that abrogates apoptosis and acts in the lower portion of the crypt. The genetics of the various subsequent steps appear to be heterogeneous. The genetic heterogeneity may arise because, in particular cases, hypermethylation knocks out different DNA repair genes (Jass et al. 2002a). Elevated somatic mutation rate for a particular spectrum of genes follows, the particular spectrum depending on the DNA repair system reduced by methylation. Increased somatic mutation can lead to rapid progression from dysplastic serrated adenomas to carcinomas.

A high MSI pathway may begin after methylation and suppression of the MMR gene *hMLH1*. By contrast, a low MSI pathway may follow after promoter methylation and suppression of the DNA repair gene *MGMT*. The enzyme MGMT removes promutagenic adducts from guanine nucleotides. Several common carcinogens create such adducts, typically in the distal colon and rectum. Loss of MGMT probably increases the rate of certain types of mutations, leading to a particular spectrum of mutated genes in subsequent progression.

SUMMARY

I described four pathways to colorectal cancer: mismatch repair mutations leading to microsatellite instability, HNPCC, hypermethylation with high microsatellite instability, and hypermethylation with low microsatellite instability. I emphasized the details because colorectal cancer provides the greatest insight into multistage progression of disease. The different pathways highlight the need to classify disease by pathway rather than solely by tissue location. In particular, the various pathways have different stages and rates of transition between stages.

In the future, it may be possible to couple better understanding of distinct colorectal pathways with measurement of age-onset patterns for each pathway. Of course, we will never have all the genetic details or perfect measurement of age-onset patterns. But we should be able to formulate and test comparative hypotheses: pathways with fewer rate-limiting stages or faster transitions between stages will differ predictably in age-onset patterns when compared with pathways that have more stages or slower rates of transition. In the next chapter, I discuss the great importance of formulating and testing comparative hypotheses. For now, I end this section by briefly summarizing the four colorectal pathways that I have discussed.

First, initiation of the classical pathway usually requires mutation of *APC* or *β-catenin,* leading to dysplastic crypt foci. Further mutations lead to adenomas, a slow transition to carcinomas, and about a 30:1 ratio of adenomas to carcinomas. Chromosomal instability, loss of heterozygosity, and aneuploidy occur. The classical pathway accounts for the majority of colorectal cancers.

Second, inherited mutations to mismatch repair (MMR) cause hereditary nonpolyposis colorectal cancer (HNPCC). This disease follows the same morphological stages as the classical pathway, but with different mutations and rates of progression. Mutations usually occur in repeated regions of genes, because reduced MMR causes increased frameshift mutations in repeated sequences. Progression through the middle stages occurs more rapidly than in the classical pathway, reducing the adenoma to carcinoma ratio to about 1:1. The HNPCC pathway lacks chromosomal instability, instead using the malfunction in DNA repair to raise the mutation rate. This pathway accounts for only about 2–4 percent of colorectal cancers.

Third, hypermethylation silences MMR, causing a high somatic mutation rate in repeated sequences. The morphological pathway and the set of mutated genes differ from HNPCC, even though both pathways have MMR defects. In this hypermethylation pathway, the initiating stages that abrogate apoptosis may focus on regulatory systems other than APC and β-catenin. Morphologically, initiation leads to hyperplastic crypts, followed by dysplastic outgrowths from these aberrant crypt foci. Subsequent mutations and gene silencing depend both on changes to repeated DNA sequences and on methylation and silencing of other genes. After initiation and progression through the early dysplastic adenoma stage, progression may be rapid, causing a low adenoma to carcinoma ratio. As in HNPCC, this sequence lacks chromosomal instability. About 10–15 percent of colorectal cancers follow this pathway.

Fourth, hypermethylation may silence DNA repair systems other than MMR. The characteristics of progression roughly follow those in the third pathway with loss of MMR by methylation. However, the particular type of DNA repair affected determines the particular genes subsequently mutated during progression. Jass et al. (2002b) have argued that perhaps 20 percent of colorectal cancers follow these various routes of progression. However, supporting data remain weaker for this pathway than for the previous three.

3.5 Changes during Progression

Multistage progression simply means that transformation to cancer does not happen in a single step. That vague definition leaves open what actually happens. In the next three sections, I briefly outline some of the details.

My ultimate goal is to formulate and test hypotheses about the processes that shape quantitative aspects of cancer incidence. I will show in the following chapters that, in the absence of knowing everything that affects progression, we can still learn a great deal if we formulate and test hypotheses in the proper way. For now, I give a brief abstract of the changes that occur during progression. Those facts help to guide the formulation of appropriate hypotheses and tests.

3.6 What Physical Changes Drive Progression?

Genetic changes alter the DNA sequence composition of the genome: mutation changes a few bases; loss of a chromosome followed by duplication of the remaining homologous copy causes loss of heterozygosity; changes in chromosome numbers alter the number of gene copies; genomic rearrangements cause loss of genes or altered gene expression; and epigenetic changes in methylation or chromatin structure also affect gene expression.

Altered cells often change the signals they provide to other cells, leading to changes in gene expression, level of tissue differentiation, and regulation of tissue growth. Changes in gene expression and tissue dynamics may lead to further changes in intercellular signaling and cause successive loss of growth regulation.

Expanding tumors must acquire resources to fuel growth. This demand for resources requires enhanced blood supply and an enriched supporting connective-tissue framework, the tumor stroma (Mueller and Fusenig 2004). Tumor growth depends on signaling between the cells in the growing tumor and the complex, supporting stromal tissue.

Most tumors acquire many mutations and genomic alterations. Do those genetic changes drive tumor progression, or are those genetic changes a consequence of other processes that drive rapid mitoses and tumorigenesis?

Several lines of evidence suggest that genetic changes drive cancer progression (Vogelstein and Kinzler 2002). Inherited mutations lead to cancer syndromes that often mimic sporadic (noninherited) cancers. The inherited cases develop at a faster pace, consistent with the hypothesis that pre-existing genetic alterations bypass normally rate-limiting events in progression. In sporadic cases, certain genetic changes recur in different individuals with the same tumor type. Genetic changes sometimes happen in a more or less consistent order.

Because cancer arises in diverse ways, there will always be some exceptions to the central role of genetic change—cancer is the breakdown of normal regulatory controls, and there are many pathways by which complex regulation can fail. To show that alternatives to genetic change play a primary role, one must formulate and test quantitative hypotheses about how those nongenetic changes alter age-specific incidence.

3.7 What Processes Change during Progression?

I maintain my focus on rate processes that limit progression and influence age-specific incidence. However, we do not know exactly which processes play key roles in the dynamics of progression, and different cancers vary widely in their characteristics. So, I will provide a sample of potential issues to set the stage for formulating quantitative hypotheses in later chapters. I emphasize processes that influence cellular birth and death, processes that generate variation in cells and tissues, and processes that select the successful tumor variants (Hanahan and Weinberg 2000).

ANTI-APOPTOSIS AND ABROGATION OF PROGRAMMED CELL DEATH

Cells kill themselves when they cannot repair genetic damage, when they do not receive tissue-specific survival signals that match their own cell type, or when they receive death signals from immune cells (Kroemer 2004). Cellular suicide—apoptosis and alternative pathways of programmed cell death—protects tissues from uncontrolled growth. Genetic changes that abrogate the normal cell death response commonly occur in tumors.

HYPERMUTATION AND CHROMOSOMAL INSTABILITY

Cells in most tumors have widespread genomic changes in chromosome number and arrangement (Rajagopalan et al. 2003). Those changes often arise from increases in double-strand DNA breaks or failure to repair such breaks, causing chromosomal instability. Tumors that have lost particular DNA repair pathways may have many mutations of the particular kind normally fixed by the lost repair system.

DNA repair systems monitor genetic damage. Detected damage induces repair or apoptosis: cell death and DNA repair are intimately associated (Bernstein et al. 2002). Increased mutation or chromosomal instability often first requires abrogation of apoptosis; otherwise, genetic damage leads to cell death rather than the accumulation of genetic change.

Rapid genetic change can increase the rate of progression. Some people have argued that cancer development requires such acceleration of progression (Loeb 1991). Others argue that normal rates of somatic mutation are sufficient to explain progression, and widespread genetic changes arise late in progression as a consequence of excessive cell division or other processes (Tomlinson et al. 1996).

CELL-CYCLE CHECKPOINTS AND ACCELERATORS

Cell-cycle checkpoints block progress through the cell cycle in the absence of appropriate external growth signals or in response to internal damage (Kastan and Bartek 2004; Lowe et al. 2004). These brakes on cell division often fall in the class of tumor suppressors—genes with products that can suppress uncontrolled cell division. Mutation of the tumor suppressor genes may set key rate-limiting steps in progression. Usually, both alleles of a tumor suppressor locus must be knocked out to release the brake, because the protein product from one functional copy is sufficient to keep the cell cycle in check. For example, the retinoblastoma protein blocks transition into the S phase of the cell cycle, during which the cell copies its DNA in preparation for splitting into two daughter cells (Fearon 2002). Only a proper combination of other cell-cycle controls can release the retinoblastoma block, providing a check that the cell is ready for the complex process of DNA replication.

Tumor suppressors brake cellular proliferation. By contrast, oncogenes stimulate cell division (Park 2002). For example, nondividing cells express little of the *myc* gene (Pelengaris et al. 2002). When such cells receive external growth signals, they quickly ramp up expression of *myc*, which in turn stimulates expression of many growth-related factors. Tumors often express high levels of the *myc* gene or similar oncogenes, causing rapid growth even in the absence of normally required external growth signals.

AVOIDING CELLULAR SENESCENCE

Most cells can divide only a limited number of times (Mathon and Lloyd 2001). With each cell division, the chromosome ends (telomeres) shorten because they are not copied by the normal DNA replication enzymes. After forty or so divisions, the special telomeric caps have worn down. Normal cells will not continue to divide. If cell division continues, the worn chromosome ends cause double-strand DNA breaks, leading to chromosomal rearrangements and genomic instability (Feldser et al. 2003).

Certain cells must divide many times without wearing out: the germ cells continue on without decay; the stem cells that replenish renewing epithelial tissues divide hundreds or perhaps thousands of times over the normal human lifespan. Those cells express a special enzyme, telomerase, that regenerates the full telomere during each replication cycle.

Late-stage cancer cells usually express telomerase (Mathon and Lloyd 2001). Telomerase expression may occur because the original cells that began progression were specialized to avoid senescence. Or the cancer cell lineage may have turned on telomerase during some stage of progression. If telomerase is off during early progression, the cancer cell lineage may develop frayed telomeres and genomic instability (Feldser et al. 2003). That instability creates genetic variability, perhaps enhancing the opportunity to develop a more aggressive genotype. However, the widespread chromosomal aberrations must eventually be controlled in the cancer cell lineage by turning on expression of telomerase, otherwise the lineage would probably self-destruct from genetic defects (Frank and Nowak 2004).

RESOURCE ACQUISITION AND STROMAL ECOLOGY

Progression follows in part from genetic changes that cause loss of control over cellular birth and death. But tumorigenesis is more complex than just transforming particular cells by genetic change. For example, a solid tumor cannot grow beyond 1–2mm without obtaining a blood supply. Tumor cells acquire vasculature by angiogenesis, the process of stimulating blood vessel growth through a tissue (Folkman 2002). Complex regulatory processes control angiogenesis (Folkman 2003). In the default state, blood vessels usually will not grow through the tissue of a developing tumor. To progress, the tumor must overcome angiogenic repression and stimulate the growth of a blood supply.

Signals that stimulate angiogenesis may come directly from the tumor cells or by collaboration with the complex mixture of other cell types in and around the developing tumor. Those other cells usually include fibroblasts, immune cells, and blood-vessel cells, together forming the stroma (Mueller and Fusenig 2004). Signaling between tumor and stromal cells regulates many aspects of tissue growth and differentiation. Progressive changes in tumor cells lead to secretion of various stromal-modifying signals, often disrupting tissue homeostasis in a way that mimics wound healing with enhanced angiogenesis, inflammatory response, and activation of nearby cells to secrete additional growth factors (Mueller and Fusenig 2004; Hu et al. 2005; Rubin 2005; Smalley et al. 2005).

The extracellular matrix provides another barrier to tumor expansion (Hotary et al. 2003; Yamada 2003). A network of protein and proteoglycan fibers forms a three-dimensional supporting mesh through most solid tissues. That matrix helps to keep spatial order among the cells and to limit uncontrolled expansion of a cellular clone. Developing tumors and their nearby stroma frequently secrete proteases that break down the extracellular matrix, disrupting tissue organization and providing an opportunity for clonal expansion of tumor cells (van Kempen et al. 2003).

INVASIVENESS AND NEGLECT OF DEATH SIGNALS

Some tumors invade nearby tissues or migrate to distant sites. In epithelial progression, tumor cells begin to move by breaking through the basement membrane (Liotta and Kohn 2001). That membrane walls off the epithelial layer from neighboring tissues. Tumor cells break the basement membrane by secreting proteases and changing their cell adhesion properties.

Distant migration requires transport through the blood or lymph systems. Most cells die during migration because they require the specific signals of their native tissue to avoid triggering their apoptotic response. To migrate successfully, cells must evolve to ignore this default death response (Fidler 2003; Douma et al. 2004).

Few migrating cells survive and grow in foreign tissue. But tumors send many colonists, and a few may succeed. To survive and grow in foreign tissue, the colonists must avoid defenses that normally kill foreign cells, avoid repressive anti-growth signals, and acquire resources. Migrating tumor cells often have high mutation rates or rapid genomic changes, which may help them to adapt to the new conditions required for growth.

3.8 How Do Changes Accumulate in Cell Lineages?

My goal is not to describe all changes. Rather, I seek alternative hypotheses about how the major, rate-limiting steps accumulate. Three possibilities seem most promising as points of departure for further study.

SINGLE DOMINANT CELL LINEAGE

Suppose a single original cell suffers the primary change. That cell may, for example, obtain a mutation that weakens its apoptotic response. Subsequent rate-limiting changes accumulate in the descendant lineages of that original cell. The progressing lineage creates changes in nearby tissues by signaling. At several stages, the dominant lineage expands into a precancerous population of cells; a new change then hits one of those cells, which subsequently expands and becomes the dominant lineage.

A single, continuous line of descent can be described most easily by the shape of cellular lineages in the historical pattern of cellular ancestry. In evolutionary biology, the historical pattern of ancestry is called the phylogeny or phylogenetic tree. Figure 3.5 shows an example of how phylogenetic shape corresponds to the history of accumulated changes in progression. The description ultimately reduces to the time to the most recent common ancestor among extant tumor cells. This coalescence time describes the degree to which one or a few lineages have dominated. In Figure 3.5d, with a single dominant lineage, the time to the most recent common ancestor of all extant cells is short. By contrast, in Figure 3.5a, with no dominant lineage, the coalescence time to a common ancestor is relatively long.

In precancerous colon crypts, the cells in the whole crypt often derive from a recent common ancestor: a single stem cell lineage and its descendants dominate the crypt (Kim and Shibata 2002). At any time, a few stem cells may be present. Over time, one of the stem lineages survives and the others drop out. The different stem lineages may compete, or differential success may just be a random process in which one lineage, by chance, takes over. With each replacement, the primary stem lineage

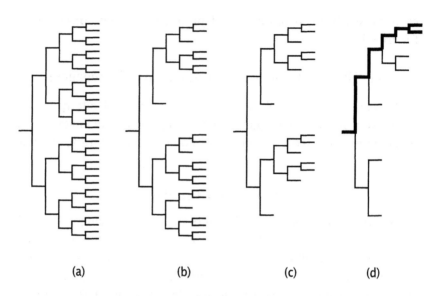

(a)   (b)   (c)   (d)

Figure 3.5   Differences in success between lineages in a phylogeny influence the shape of the tree. All trees shown with the ancestral cell of origin on the left. Time increases from left to right. (a) Shape when all lineages survive. (b) Tips that stop in the middle of the tree represent lineages that have gone extinct. Some extinctions occur in this case, but many different lineages have survived to the present. (c) Greater differential success between lineages; however, no single winner emerges in any time period. (d) Only a single lineage survives over time, shown in bold. In each time period, a single lineage gives rise to all survivors a few generations into the future. If major changes in progression cause subsequent clonal expansions, each clonal expansion arising from a particular cell, then the phylogeny will be dominated by a single lineage as in (d).

may then split off by seeding a few new stem lineages. The cycle of coalescence and splitting of lineages repeats. If early genetic changes in cancer progression do not alter the normal pattern of cellular lineages, then such changes accumulate in a dominant cell lineage of a crypt. A different cell lineage usually dominates in each crypt (Kim and Shibata 2004). A tumor usually arises from a single crypt, so a single lineage dominates early tumor evolution.

Only a few studies provide indirect information on cell lineages in a growing primary tumor. Leukemias have been analyzed more than solid tumors, because one can easily sample over time the evolving cells in the blood. Among later-stage leukemias, only a small fraction of cancer cells have the ability to regenerate a tumor (Reya et al. 2001). These cancer stem cells may form the main long-term line of tumor evolution. Some evidence suggests that cancer stem cells also occur in solid tumors (Singh et al. 2004; Dean et al. 2005). Phylogenetic analyses will eventually provide a clearer picture of cell lineage history and evolution in tumors.

Multiple Lineages

Figure 3.5c shows the ancestry splitting into two groups soon after the initial change that started progression. Two or more distinct lineages could occur if the different lineages followed independent pathways in progression, and the cells from the distinct lineages did not compete directly. Alternatively, the two lineages may provide synergistic stimulation in progression; for example, each lineage could provide complementary growth signals to its partner.

Distinct lineages may also arise independently, for example, one mutation originating in a stromal cell and a second mutation originating in an epithelial cell. Synergistic signaling between the progressive stromal and epithelial lines could play an important role in some cases. Mueller and Fusenig (2004) review several examples in which genetic changes in stromal cells appear to play a key role in progression. See Kim et al. (2006) for a recent demonstration of how progression in gastrointestinal epithelial tissue depends on interactions with stromal cells.

Phylogenetic Position of Migrant Cells and Metastases

Consider two contrasting patterns. Migrant cells may arise only from the dominant cell lineage in late-stage localized tumors. In this case, different colonists and the primary tumor would have a common cellular ancestor a short time back. Alternatively, migrant cells may arise at different stages of tumor development or from different lineages in late-stage tumors. In this case, the time back to common ancestors for colonist cells and the cells in the primary tumor would be variable; metastases derived from colonists would be genetically heterogeneous. Other phylogenetic patterns are possible: for example, a cancer stem cell lineage in the primary tumor may be numerically rare but nonetheless be the progenitor of both local and distant cell lines.

Although much has been written about which cells give rise to metastases, few data exist with regard to lineage history (e.g., Bonsing et al.

2000; Weiss 2000). Recent technological advances should make it possible to get more genomic data on various tumor and metastatic cells, so perhaps phylogenetic analyses will be available in the future.

3.9 Summary

This chapter presented evidence that cancer progresses through multiple stages. To connect those biological details on multistage progression to quantitative theories of cancer incidence, we need a way to measure the shifts in incidence caused by particular genetic and physiological changes. The early history of multistage theory provided such a connection between genetics and incidence; however, some of the insights of those early studies have been lost amid the great recent progress in genetics and biochemistry. The next chapter reviews the history of multistage theory to set the background for later chapters, in which I build the tools needed to develop quantitative analyses of the causes of cancer.

4 History of Theories

In this chapter, I discuss the history of theories of cancer incidence. I focus only on those aspects of history that remain relevant for current research on progression dynamics and incidence. More details about the history and the literature can be obtained from the many published articles that review theories of cancer incidence (Armitage and Doll 1961; Druckrey 1967; Ashley 1969b; Cook et al. 1969; Doll 1971; Nowell 1976; Peto 1977; Cairns 1978; Whittemore and Keller 1978; Scherer and Emmelot 1979; Moolgavkar and Knudson 1981; Forbes and Gibberd 1984; Stein 1991; Tan 1991; Knudson 1993, 2001; Lawley 1994; Iversen 1995; Klein 1998; Michor et al. 2004; Moolgavkar 2004; Beckman and Loeb 2005).

The first section introduces the original theories of multistage progression. Starting in the 1920s, several experimental programs applied chemical carcinogens to animals. Two different carcinogens applied in sequence often yielded a higher rate of cancer than did application of a single carcinogen over the same time period. This synergistic effect between two carcinogens led to the idea that each carcinogen stimulated a different stage in progression: the two-stage model of carcinogenesis.

A separate line of multistage theories began in the 1950s by analysis of the observed rates of cancer at different ages. For most of the common adult cancers, the age-specific incidence curves rise with a high power of age, roughly proportional to t^{n-1}, where t is age and, from the data, $n \approx 6$. Mathematical models showed that such incidence curves would occur if cancer follows after progression through n rate-limiting steps. This analysis led to the hypothesis of multistage progression.

The second section turns to the most profound empirical tests of multistage theory. The mathematical theory predicted that the greater the number of rate-limiting steps, n, the faster incidence rises with age. Ashley (1969a) and Knudson (1971) reasoned that if somatic mutation is the normal cause of progression, then individuals who inherited a mutation would have one less step to pass before cancer develops. Multistage theory makes the following prediction: inherited cases with a smaller number of steps to pass have a slower rise of incidence with age than noninherited cases. Data comparing inherited and noninherited cases in colon cancer (Ashley 1969a) and retinoblastoma (Knudson 1971) supported this prediction.

The third section takes up the kinds of changes that cause progression. Many authors have emphasized genetic changes by somatic mutation. However, critics have argued against the somatic mutation theory, favoring instead alternative mechanisms of genomic and physiological change. For understanding the kinetics of progression, the alternative mechanisms of change set different constraints on the rate parameters of progression but do not alter the basic understanding of multistage theory.

The fourth section highlights a puzzle about somatic mutation rates and progression. Commonly cited values for the normal rate of somatic mutation typically fall near 10^{-6} mutations per gene per cell division. Mutations to six particular genes in a cell lineage would occur with probability 10^{-36} multiplied by the number of cell divisions in that lineage. Historically, calculations of this sort with various assumptions about the number of cell divisions and the number of cells at risk have suggested that normal somatic mutation does not occur fast enough to explain observed cancer incidence by progression through numerous stages. That conclusion has led to various alternative theories about hypermutation, selection, clonal expansion of precancerous cell lineages, and fewer numbers of mutations required for progression.

The fifth section reviews the theory of clonal expansion. Suppose a mutation arises in a cell and that cell proliferates into a large clone. The probability of a second mutation in a cell rises as the number of target cells carrying the first mutation increases. Thus, clonal expansion can greatly increase the rate at which mutations accumulate in cell lineages.

The sixth section continues discussion of cell lineages and mutation accumulation. The rate at which cells divide is important because mutations happen mostly during cell division. Tissues that grow early in life and then slow to a very low rate of cell division predominantly suffer childhood cancers rather than adult cancers. By contrast, epithelial tissues with continual cell division throughout life suffer mostly adult cancers and account for about 90% of human cancers. Cairns (1975) emphasized that certain epithelial tissues renew from stem cells, a tissue architecture that greatly reduces competition between lineages and reduces opportunities for clonal expansion. Without clonal expansion, mutations must arise solely within a lineage of single cells.

The seventh section follows with theories for how multiple mutations accumulate in cell lineages. Some authors emphasize hypermutation, in which an early step of carcinogenesis reduces DNA repair efficacy or promotes chromosomal aberrations during cell division. Once the caretakers of genomic integrity have been damaged, subsequent changes may accumulate relatively rapidly. Other authors emphasize competition between genetically variant cell lineages. Such selection between variants favors clonal expansion of more aggressive cell lines. Tissue architectures that reduce cell lineage competition provide some protection against cancer.

The eighth section extends the topic of the mutation rate. I mentioned that, with regard to kinetics, any heritable genomic change that alters gene expression can influence cancer progression. Recent work on epigenetic processes shows that heritable genomic changes often accumulate by DNA methylation and histone modification. Tumors frequently have elevated rates of epigenetic change, providing another pathway to increase the rate of progression.

4.1 Origins of Multistage Theory

Two different lines of thought developed the idea that cancer progresses through multiple stages. The first line arose from the observation that, in experimental animal studies, cancer often followed after sequential application of different chemical carcinogens. The second line arose from observations on the age-onset patterns of cancer, in which incidence often accelerates with age in a manner that suggests multiple stages in progression.

EXPERIMENTAL CARCINOGENESIS

In the 1920s, several laboratories began to apply chemical carcinogens to experimental animals. Deelman (1927) summarized observations in which repeated applications of tar to skin led to a small number of tumors, after which tarring was stopped. A few days later, the skin was cut where no tumors had appeared. Most incisions developed tumors in the scars; most such tumors were very malignant. Two distinct processes, tarring and wounding, combined to cause aggressive cancers.

Twort and Twort (1928, 1939) described several experimental protocols in which sequential application of different chemicals was much more carcinogenic than either agent alone. In the early 1940s, several others, notably Rous and Berenblum, reported similar observations on the co-carcinogenic interaction between two different treatments when applied sequentially (MacKenzie and Rous 1940; Berenblum 1941; Rous and Kidd 1941).

Friedewald and Rous (1944) described the first treatment as an *initiator*, because it seemed to initiate the carcinogenic process but was usually not sufficient by itself to cause cancer. They called the second treatment a *promoter*, because it caused progression of previously initiated cells but by itself rarely led to cancer. In a series of papers, Berenblum and Shubik (1947b, 1947a, 1949) synthesized the experimental studies and thinking on co-carcinogenesis into the two-stage theory of initiation and promotion.

The mechanistic action of initiators and promoters has been widely debated. In some cases, it was thought that the initiator is mutagenic, causing latent DNA lesions in some cells, and the promoter is mitogenic, stimulating cell division and providing favorable conditions for tumor formation. However, no simple mechanistic explanation fits all cases. Indeed, many observations from experimental carcinogenesis do not fit with a simple two-stage explanation (Iversen 1995).

The initial theory provided a useful framework for the early experimental studies, but hardened too much into "two-stage" and "initiator-promoter" slogans that probably hindered as much as helped to understand the actual mechanisms of carcinogenesis (Iversen 1995). Recent emphasis has moved closer to the actual molecular mechanisms involved, aided by the great technical advances now underway. Aspects of initiation and promotion may play a role, but the older dominance of the rigid two-stage theory has naturally faded. For our purposes, the two-stage theory is important because it provided the first evidence and thinking with regard to multiple stages in cancer progression.

AGE-SPECIFIC INCIDENCE

Two observations about cancer incidence in epithelial tissues have led to multistage theories. First, cancer incidence often increases rapidly with age. Second, what happens to any particular individual appears to be highly stochastic, yet simple patterns emerge at the population level.

In a rarely cited paper, Charles and Luce-Clausen (1942) developed what may be the first quantitative multistage theory. They analyzed observations on skin tumors from mice painted repeatedly with benzopyrene. They assumed that benzopyrene causes a mutation rate, u, and that cancer arises by knockout of a single gene following two mutations, one to each of the two alleles. If t is the time since the start of painting with the carcinogen, then the probability of mutation to a single allele is roughly ut, and the probability of two hits to a cell is $(ut)^2$. They assumed that painting affects N cells, so that $N(ut)^2$ cells are transformed, and that the time between the second genetic hit and growth of the transformed cell into an observable papilloma is i.

From these assumptions, the number of tumors per mouse after the time of first treatment is $n = N[u(t - i)]^2$. This formula gave a good fit to the data with reasonable values for the parameters. Thus, Charles and Luce-Clausen (1942) provided a clearly formulated multistage theory based on two genetic mutations to a single locus and fit the theory to the age-specific incidence of tumors in a population of individuals. They assumed that both genetic hits must happen to a single cell, after which the single transformed cell grows into a tumor.

Muller (1951, p. 131) mentioned the need for multiple genetic hits: "There are, however, reasons for inferring that many or most cancerous growths would require a series of mutations in order for the cells to depart sufficiently from the normal." However, Muller did not connect his statement about multiple genetic hits to age-specific incidence.

The next theoretical developments followed directly from the observation that several cancers increase in incidence roughly with a power of age, t^{n-1}, where t is age and the theories suggested that n is the number of rate-limiting carcinogenic events required for transformation. Fitting the data yielded $n \approx 6\text{-}7$ distinct events.

Whittemore and Keller (1978) usefully separate explanations for the exponential increase of incidence with age between multicell and multistage theories.

The multicell theory assumes that the distinct carcinogenic events happen to $n \approx 6\text{-}7$ different cells in a tissue (Fisher and Hollomon 1951). If the carcinogenic events occur independently in the different cells, then this process would yield an age-specific incidence proportional to t^{n-1}, matching the observations. In particular, this theory leads to an expected incidence of

$$I(t) \approx (Nu)^n t^{n-1} / (n-1)!, \tag{4.1}$$

where N is the number of cells at risk for transformation, and u is the transformation rate per cell per unit time; thus, Nu is the rate at which each transforming step occurs in the tissue.

The multistage theory assumes that changes to a tissue happen sequentially. Charles and Luce-Clausen (1942) explicitly discussed and analyzed quantitatively two sequential mutations to a particular cell; Muller (1951) discussed in a general way sequential accumulation of mutations. Nordling (1953) introduced log-log plots of incidence data to infer the number of steps. Nordling (1953) assumed that the steps were sequential mutations to a cell lineage, and he suggested that a log-log slope of $n - 1$ implied n mutational steps in carcinogenesis. From data aggregated over various types of cancer, he inferred $n \approx 7$.

Stocks (1953) followed Nordling (1953) with a mathematical analysis to show how sequential accumulation of n changes to a cell leads to log-log incidence plots with a slope of $n - 1$. Stocks (1953) had the right idea, although from a mathematical point of view his analysis was rather limited because he assumed that changes happened at a constant rate per year and that at most one change per year occurred.

Armitage and Doll (1954) crystallized multistage theory by extending the data analysis and mathematical development. With regard to the data, they examined log-log plots for several distinct cancers rather than aggregating data over different cancers as had been done by Nordling (1953). With regard to theory, their mathematical model allowed different rates for different steps; they assumed continuous change rather than arbitrarily limiting changes to one per year; and they noted that the stages did not have to be genetic mutations but only had to be sequential changes to cells. The style of data analysis and mathematical argument formed the basis for the future development of multistage models.

Armitage and Doll (1954) rejected Fisher and Hollomon's (1951) multicell theory in which the changes happen to different cells. Armitage and Doll argued that if a chemical mutagen caused cancer by causing mutations to several different cells, then incidence would increase with dose raised to a high power. For example, in Eq. (4.1), if the mutation rate, u, increases linearly with dose, d, then for n steps in carcinogenesis, the incidence is proportional to d^n. In those cases known to Armitage and Doll, incidence increased only with a low power of dose but a high power of time. Thus, they rejected the multicell theory.

Against Armitage and Doll's quick rejection of multicell theory, Whittemore and Keller (1978) pointed out that if a particular carcinogen affected only a few of the various stages in progression, for example only $m < n$ of the stages, then multicell theory predicts that incidence would increase as d^m. So, Armitage and Doll's argument did not really rule out the multicell theory. Later molecular evidence tends to favor sequential changes to a cell lineage rather than changes to many different cells. However, recent work on genetic changes in stromal cells and analyses of the tissue environment (see below) will probably lead to the conclusion that changes to the surrounding cells and tissue can also be important in some cases.

The next step in the history, from a chronological point of view, concerns the role of cell proliferation and clonal expansion. However, I delay that topic until a later section. Instead, I take up what I consider to be the next major insight: how to test theories of progression.

4.2 A Way to Test Multistage Models

Various forms of multistage theory can be fit to the data. But the fact that a particular model can be fit to the data by itself provides only weak support for the model. The problem is that models are often too pliable, too easily fit to different forms of data. Because many different models can be nicely fit to the same data, fitting models to data provides very little insight. For testing multistage hypotheses, the key breakthrough came with Knudson's (1971) comparison of incidence between inherited and noninherited forms of retinoblastoma. In this section, I present the background to Knudson's work, what he accomplished in his studies, and some of the historical aspects of his work (Knudson 1977).

In the 1960s, the importance of somatic mutations and the nature of stages in progression continued to be debated (Foulds 1969). Several authors developed the idea that cancer arises by the accumulation of genetic mutations to cell lineages. Burch (1963) noted that if a sequence of mutations drives progression, then some individuals may inherit one mutation and obtain the rest after birth by somatic mutation. Burch

(1964) stated: "Although for a specific cancer the inherited predisposition usually affects only a single autosomal locus ... the phenotypic expression in adults should generally involve somatic mutation of the gene homologous with the inherited allele, together with somatic mutation of homologous genes at another locus." This combination of inherited and somatic mutation explains why the "commonest form of predisposing inheritance appears to be a simple Mendelian dominant of incomplete penetrance."

Anderson (1970) summarized further evidence of autosomal dominant inheritance of cancer predisposition for certain types of tumors, including retinoblastoma. In discussion of Anderson's paper, DeMars (1970) stated:

> I think many pedigrees are consistent with the notion that one of the parents in these families might be heterozygous for a recessive and that the neoplasms appear as a result of subsequent somatic mutations in which individual cells become homozygous for a recessive neoplasm-causing gene. Can you critically exclude that possibility in any of the cases that you called autosomal dominant? It's obviously important if we want to understand the relationship between the genotypes and the phenotype called cancer.

Ashley (1969a) made the first comparison of age-specific incidence between inherited and noninherited forms of the same cancer. He compared polyposis coli, an inherited form of colon cancer, with noninherited cases. He concluded that "the slope of age dependence for the development of colonic cancer is less steep in the case of individuals carrying the gene for polyposis coli than in the general population." Ashley argued that this comparison supported multistage theory, where transitions between stages arise by genetic mutations (hits): "the difference in slopes suggests that more 'hits' are required in the case of an individual in the general population before a colonic cancer will develop than is the case in an individual who has, in his genome, the gene [mutation] for polyposis coli."

Knudson (1971) compared age-onset patterns of retinoblastoma between inherited and noninherited forms. In his introduction, Knudson placed his work in the context of multistage progression, in which progression is driven by genetic mutations:

The origin of cancer by a process that involves more than one discreet [sic] stage is supported by experimental, clinical, and epidemiological observations. These stages are, in turn, attributed by many investigators to somatic mutations ... What is lacking, however, is direct evidence that cancer can ever arise in as few as two steps and that each step can occur at a rate that is compatible with accepted values for mutation rates. Data are presented herein in support of the hypothesis that at least one cancer (the retinoblastoma observed in children) is caused by two mutational events.

Knudson concluded from his retinoblastoma data that individuals who inherit one mutation follow the age-onset pattern expected if one additional hit leads to cancer, whereas individuals without an inherited mutation follow the kinetics expected if two hits leads to cancer. Knudson fit his data to particular one-hit and two-hit mathematical models. However, his theoretical arguments in this paper ignored the way the retina actually develops. In a later pair of papers, Knudson and his colleagues produced a theory of incidence that accounts for retinal development (Knudson et al. 1975; Hethcote and Knudson 1978).

The later papers had several parameters concerning retinal development and mutation that the authors fit to the data. However, Knudson's great insight was simply that age-specific incidence of inherited and noninherited retinoblastoma should differ in a characteristic way if cancer arises by two hits to the same cell. He obtained the data and showed that very simple differences in incidence do occur.

In my view, nothing is more powerful than figuring out how to test an important hypothesis by a simple comparison (Frank 2005; Frank et al. 2005). Although Ashley (1969a) made essentially the same comparison of age-specific incidence between inherited and noninherited forms of colon cancer, Knudson's (1971) work achieved the status of a classic whereas Ashley's (1969a) paper is rarely cited. Ashley certainly deserves credit for his accomplishment, but Knudson's paper deserves to be regarded among the few major achievements in this subject.

In retrospect, we can now see that Knudson's paper made two major contributions. First, he compared age-specific incidence curves between inherited and noninherited cases. The inherited cases had increased incidence by an amount consistent with an advance of progression by one rate-limiting step. This approach provided a method of analysis by which one could use quantitative comparison of age-specific incidence between two groups to infer underlying processes of progression. In this case, the comparison pointed to a genetic mutation as a key rate-limiting step.

The second contribution arose from the hypothesis that two mutations provide the only rate-limiting barriers to tumor progression in retinoblastoma. Knudson's conclusion that two genetic hits lead to cancer contributed an important step in the history of the subject. In particular, Knudson's study presented the first data in support of the idea that cancer is primarily a genetic disease driven by mutation and that progression can be explained by known rates of mutation. Later, when it was discovered that the genetic basis of retinoblastoma depended on mutational knockout of both alleles at a single locus—named the retinoblastoma or *Rb* locus—Knudson's hypothesis provided the link between the rate of cancer progression and the molecular nature of tumor suppressor genes, in which abrogation by mutation of both alleles knocks out the function of a tumor suppressor protein and releases a constraint on tumorigenesis.

Knudson (1971) has been cited 2,926 times as of August, 2005. Figure 4.1 shows the citation history by year. The sharp increase in citations in the early 1990s follows the rise of molecular studies that confirmed the key role of tumor suppressor genes in limiting cancer progression and the contribution of mutations to tumor suppressor genes in tumorigenesis (Knudson 2003).

A dissonance exists between Knudson's quantitative method of analysis, which formed the entire basis for his paper, and the molecular analyses of the 1990s that elevated Knudson's work to classic status. The later molecular work cited Knudson because he foreshadowed the conclusions of the molecular analyses: cancer progression requires knockout of both alleles of a tumor suppressor locus. But that molecular work has ignored the major intellectual contribution of the Ashley-Knudson papers: the quantitative analysis of progression dynamics by comparison of age-specific incidence curves between different genotypes.

I have emphasized several times that a gene has a causal effect on cancer to the extent that it has a quantitative effect on progression dynamics: a genetic change has a causal effect to the extent that the genetic change shifts the age-specific incidence curve. Ultimately, research must return to this quantitative problem. I develop this issue in the next chapter.

Figure 4.1   Citations per year for Knudson (1971) as of August, 2005.

4.3 Cancer Is a Genetic Disease

The role of somatic mutations in cancer was debated for many years. Witkowski (1990) puts that historical debate in context with a comprehensive time line of developments in cancer research interleaved with developments in basic genetics and molecular biology (see also Knudson 2001). Here, I mention a few of the highlights that provide background for evaluating theories of progression and incidence.

Boveri (1914, 1929) often gets credit for the first comprehensive theory of somatic genetic changes in cancer progression (Wunderlich 2002). Tyzzer (1916) used the term "somatic mutation" to describe events in cancer progression. In the 1950s, Armitage and Doll (1954, 1957) cautiously described the stages of multistage progression as possibly resulting from somatic mutations but perhaps arising from other causes. Burdette (1955), in a comprehensive review of the role of genetic mutations in carcinogenesis, tended to oppose the central role of mutations in progression. In (1969), Fould's extensive summary of cancer progression also downplayed the role of mutation.

Knudson's (1971) study strongly supported mutation as the primary cause of progression. But Knudson's evidence for the role of mutation came indirectly through quantitative analysis of incidence curves; I suspect that Knudson's study had only limited impact at the time with regard to the debate about the importance of mutation.

The first steps in the modern molecular era began in the late 1970s, with the cloning of the first oncogenes that stimulate cellular proliferation. In the 1980s, several groups cloned the *Rb* (retinoblastoma) gene and other tumor suppressor genes. The tumor suppressors stop the cell cycle in response to various checkpoints (see review by Witkowski 1990). From these molecular studies arose the concept that oncogene loci require mutation to only one allele to stimulate proliferation, because the mutant allele provides an aberrant positive control, whereas tumor suppressor loci require mutations to both alleles to abrogate the negative control on the cell cycle: one hit for oncogenes, two hits for tumor suppressor genes.

Fearon and Vogelstein (1990) provided the next step with their genetic analysis of colorectal tumor progression. They isolated tumors in different morphological stages of progression. From genetic analysis of those samples, they concluded that mutational activation of oncogenes and mutational inactivation of tumor suppressor genes drive progression. Fewer genetic changes in key oncogenes and tumor suppressor genes lead to benign tumors; more changes lead to aggressive cancers. The mutations tend to happen in a certain order, but much variability occurs. Five or so key mutations seem to be involved in progression. The mutations accumulate in a cell lineage over time, leading to monoclonal tumors. Together, these observations support multistage carcinogenesis by the accumulation of mutations in cell lineages.

The initial studies of cancer genes focused on changes in progress through the cell cycle: mutations to oncogenes typically accelerated the cycle, and mutations to tumor suppressor genes typically released blocks to cell-cycle progress. Further studies showed that many cancer-related genes influence DNA repair and chromosomal homeostasis. Mutations in such genes increase the rate of point mutations, the loss of chromosomes, the accumulation of duplicate chromosomes, and several varieties of chromosomal instability. Most cancers appear to have some sort of breakdown in DNA repair capacity or in chromosomal homeostasis. Kinzler and Vogelstein (1998) named those genes that regulate the cell-cycle "gatekeeper" genes and those genes that manage genetic integrity "caretaker" genes.

A distinct line of theory focuses on the important role of tissue interactions instead of the accumulation of mutations in cell lineages. For example, Folkman (2003) emphasizes angiogenesis—the recruitment of a blood supply to a growing tumor. In developing epithelial tumors, the neighboring stromal tissue interacts in many ways with the primary growth (Mueller and Fusenig 2004).

With regard to tissue interactions, perhaps the key problem concerns the nature of rate-limiting steps in progression. For example, a primary cell lineage that is accumulating mutations and progressing toward cancer may acquire a mutation that alters the neighboring stromal tissue or attracts a blood supply. Kinzler and Vogelstein (1998) call such mutations "landscapers." Alternatively, genetic changes may arise in the neighboring tissue rather than in the primary cell lineage that has started toward tumor progression. Or changes in tissue may be limited by physiological processes that do not derive from underlying genetic changes.

In summary, the dominant view at present focuses on accumulation of genomic changes in one or perhaps a few cell lineages. Tissue interactions, such as angiogenesis and signals from the stromal environment, clearly influence tumorigenesis, but their relative importance compared to genetic change in limiting the quantitative rate of progression remains unknown. Finally, other types of genomic changes that regulate gene expression may be important, such as methylation of DNA promoter regions and modification of histones. I discuss below how such genomic changes in gene regulation may influence rates of progression.

With these modern views of mutation accumulation and cancer progression in mind, I return to the problem of mutation rates. That problem influenced the development of theoretical models.

4.4 Can Normal Somatic Mutation Rates Explain Multistage Progression?

By the 1950s, studies of age-specific incidence in humans and chemical carcinogenesis in animals supported the theory that cancer progresses through multiple stages. The first quantitative theories of Nordling (1953) and Armitage and Doll (1954) inferred approximately six stages.

Ashley (1969b) used the standard Armitage and Doll (1954) multistage model to fit data for gastric cancer. He calculated $n = 7$ stages and a mutation rate of 10^{-3}. His calculations are a bit hard to follow, but he seems to be using somatic mutation rate per year. He concluded that the fitted mutation rate appears to be high, although he seemed not to be aware of the scaling he used for his mutation rate estimate. In any case, this high number may have influenced subsequent authors by suggesting that the standard multistage theory requires a very high mutation rate. For example, Knudson (1971) stated in his introduction: "What is lacking, however, is direct evidence that cancer can ever arise in as few as two steps and that each step can occur at a rate that is compatible with accepted values for mutation rates."

Stein (1991, p. 167) provides the following calculation to support his argument that five or more hits are very unlikely based on standard somatic mutation processes:

> It is generally agreed that mutation rates in mammalian cells occur with a frequency of some 10^{-5} to 10^{-6} mutations per cell generation (Evans 1984) [see also Lichten and Haber (1989), Yuan and Keil (1990), Kohler et al. (1991)]. Thus, five independent, simultaneous mutations will occur at a frequency of some 10^{-25} to 10^{-30} mutations per cell generation. To score such a 5-hit event will require the elapse of some 10^{25} through 10^{30} generations. Now the human body, in an average lifetime, produces a total of only 10^{16} cells, or that number (minus one) of cell divisions. By this calculation, on a 5-hit model, cancer should seldom occur—indeed, in not more than 10^{-9} down to 10^{-14} of the population—that is, never. The model requires mutation rates of some 10^{-3} per cell division for it to be applicable, rates which are most unlikely to be found.

The apparent contradiction between the commonly accepted somatic mutation rate and those rates supposedly needed for a multiple-hit theory may have played an important role in how the theory developed. In particular, Loeb has emphasized that an early stage in carcinogenesis must very often be mutation to the DNA repair system (Loeb 1991; Beckman and Loeb 2005). Subsequent hypermutation could then explain how cancer cells obtain the multiple mutations that most tissues apparently need for transformation (Rajagopalan et al. 2003; Michor et al. 2004). The fact that many tumors have chromosomal instabilities supports the hypermutation theory.

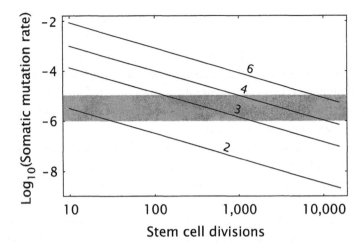

Figure 4.2   Mutation rate per cell division required to explain observed cancer incidence for various numbers of stages in multistage progression. The x axis shows the number of cell divisions over a lifetime, d. The calculations follow a simple multistage model with constant mutation rate per cell division, the same mutation rate for each transition between stages, and no clonal expansion. Cancer arises only after the accumulation of n mutations within a single cell lineage. The number attached to each line show the number of stages in progression, n, from classical multistage theory. The shaded area highlights the commonly accepted mutation rate per cell division. I calculated the required mutation rate per cell division, u, by solving for the value of u that satisfies the equation $N(1 - \sum_{i=0}^{n-1} P_i(ud)) \approx C$, where N is the number of distinct cell lineages in the tissue under study, $P_i(x)$ is the Poisson probability of i events given a mean of x events, d is the number of cell divisions per cell lineage over a lifetime in that tissue, and C is the probability that an individual develops cancer in that tissue. For this figure, I used $N = 10^8$ and $C = 0.05$; results change little when varying N up or down by a factor of 10 and when varying C over the range $0.01 - 0.1$. See Chapter 6 for the mathematical background.

The need for hypermutation seems to be widely accepted (but see Sieber et al. 2005). However, my own calculations of the somatic mutation rate required to get several hits contradicts the calculation given by Stein (1991) and the strong conclusions drawn by Loeb (1991) and Beckman and Loeb (2005) on the sheer implausibility of multiple mutations accumulating in a single cell lineage (see also Calabrese et al. 2004).

Figure 4.2 shows that a somatic mutation rate on the order of 10^{-5} to 10^{-6} may be sufficient to explain 4–6 hits. I used a model in which stem cells renew tissues, as happens in colorectal, epidermal, and perhaps several other epithelial tissues, in which most human cancers arise.

Colonic epithelium renews one to two times per week, so stem cells probably divide 50–100 times per year. Over a lifetime, the number of stem cell divisions to renew the colonic epithelium may be near 10^4. Other tissues renew less frequently, perhaps needing somewhere around 10^2 to 10^3 stem cell divisions. Figure 4.2 shows that 10^4 stem cell divisions can explain 4–6 hits within the normal range for somatic mutation; 10^3 cell divisions can explain 3–4 hits.

Hypermutation may indeed play a key role in many cases. However, in looking at Figure 4.2 and the calculations in Calabrese et al. (2004), the argument against standard mutational processes does not seem as strong as is sometimes presented.

The debate about the role of hypermutation continues in the current literature. I delay discussion of those arguments until a later section, so that I can first fill in important steps in the historical development of the subject.

4.5 Clonal Expansion of Premalignant Stages

Muller (1951, p. 131) described the problem clearly. In the accumulation of a series of somatic mutations within a cell lineage:

> The time element would constitute an influential factor unlike what is found to be the case in ordinary mutation production; for cells in which one step had occurred might because of it have proliferated sufficiently, by the time of a later treatment, to give better opportunity for another step to occur on top of the first.

Nordling (1953) made a similar comment, but, having cited Muller in another context, may well have obtained the idea from the quote here.

Platt (1955) independently came to the same idea when thinking about the long latent period between exposure to a carcinogen and occurrence of cancer. Platt argued that

> If the carcinogen simply acts by causing cells to proliferate, so that instead of dividing by mitosis x times in 20 years, they have been stimulated to divide $x \times y$ times ($y > 1$), and if, as Sonneborn seems to have shown in paramecium, the chromosomal substance duplicates more and more inaccurately as the number of divisions is increased, and if this kind of nuclear aberration could cause a malignant change in the cell, the reason for the latent period would be explained.

Armitage and Doll (1957) developed a two-stage mathematical theory in which the first hit causes proliferation of the altered cell, and the second hit causes progression to cancer. They developed this theory to explain two observations. First, prior experimental studies of carcinogens had emphasized only two distinct stages in carcinogenesis. Second, many common cancers increased in incidence with about the fifth or sixth power of age.

Previously, Armitage and Doll (1954) showed that a simple multistage model could explain the increase of incidence with age based on six or seven hits, the number of hits being the exponent on age plus one. However, given the two-stage interpretation of experimental carcinogen studies, they sought in their 1957 paper an alternative theory to fit the data. Their new clonal expansion theory could be fit to the observed rise of incidence with a high power of age. The rapid increase in incidence with age occurs because, given the first hit, the rate of transformation by the second hit increases with time as the clone of initiated cells grows and raises the number of cells at risk for obtaining the second hit.

Starting with Fisher (1958), many others have given variant mathematical treatments of clonal expansion. They all come down to the same process: increasing the number of target cells with $i - 1$ hits raises the rate at which the ith hit occurs. This increase in the rate of transition between stages raises the slope of the incidence curve (acceleration), allowing a model with a small number of hits to generate incidence curves with high acceleration.

I develop some of the technical details of clonal expansion models in the mathematical chapters of this book. For example, if the clone expands rapidly, the next hit comes so quickly that it is not rate limiting in progression. Once a clone approaches in size the inverse of the mutation rate, the next hit comes inevitably and does not limit the rate of progression. So, these models depend on slow clonal expansion over many years to provide a fit to observed incidence curves.

Recent molecular studies implicate several key genetic changes in progression for many cancers. Because of those studies, the two-stage models of clonal expansion have given way to more sophisticated multistage models that include one or more stages of clonal proliferation (Luebeck and Moolgavkar 2002).

Many models can provide a moderately good fit to the data for common cancers. Thus, the data do not strongly discriminate between the original multistage theory, the two-stage clonal expansion theory, or the newer hybrid models. Armitage and Doll's (1961, p. 36) conclusions still apply:

> In summary, we doubt whether the available observational data provide clear and consistent evidence in favor of any particular model. Further elucidation is likely to come either from direct biological evidence of a nonquantitative nature, or from quantitative experiments, carefully designed and reported, perhaps on a larger scale than is usually undertaken at present.

I agree that one cannot easily choose between the main classes of models by analyzing how well they fit the data. Most of the models supply a set of reasonable assumptions or modifications that provide a good fit. However, I do think that comparative tests like those originally used by Ashley (1969a) and Knudson (1971) can be developed to discriminate between the models (Frank 2005; Frank et al. 2005). I discuss that approach in Chapter 8.

4.6 The Geometry of Cell Lineages

Two aspects of cellular reproduction influence mutation accumulation. First, the rate of cell division influences the number of mutational events per unit time, because mutations arise primarily during cell replication. Second, the shape of cellular lineages determines how a single mutational event passes to descendant cells of a lineage. The rate at which a second hit strikes a descendant cell depends on how many of those descendant cells exist.

Some tissues have extensive cell division early in life and then relatively little after childhood, for example, neural and bone tissue. The relatively rare childhood cancers occur in such tissues, whereas the common adult cancers occur in continuously dividing tissues. Perhaps as much as 90% of human cancers arise in renewing epithelial tissues, most commonly, those of the colon, lung, breast, and prostate.

I am not certain about the historical origins of these ideas on cell division. The early chemical carcinogenesis literature emphasized the role of cell division rate stimulated by particular chemical agents. With regard to childhood cancers and tissue growth, Moolgavkar and Knudson (1981) reviewed some prior work and then presented an extensive mathematical framework in which to analyze the role of development in cell division and age-specific incidence. Moolgavkar and Knudson focused on extending the two-hit theories with clonal expansion to fit the age-incidence curves of both childhood and adult cancers.

Cairns (1975) wrote the key paper on cell lineage shape in epithelial tissues. He emphasized three factors that reduce mutation accumulation and the risk of cancer.

First, renewal of epithelial tissue from stem cells creates a linear cellular history that reduces opportunities for multiple mutations to accumulate in a lineage. Normally, each stem cell division gives rise to one stem cell that remains at the base of the epithelium and one transit cell. The transit cell divides a limited number of times, producing cells that move up from the basal layer and eventually slough off from the surface. The stem lineage renews the tissue and survives over time. Thus, accumulation of somatic mutations occurs mainly in the stem lineage. Mutations in transit cells usually are discarded as the transit cells die at the surface.

Recent studies of human epidermal tissue suggest that the skin renews from relatively slowly dividing basal stem cells that give rise to rapidly dividing transit lineages, each transit lineage undergoing 3–5 rounds of replication before sloughing from the surface (Janes et al. 2002). Studies of gastrointestinal tissues estimate 4–6 rounds of division by transit lineages (Bach et al. 2000). Sell (2004) reviews the nature of stem cells in other tissues.

Second, stem cells may have reduced mutation rates compared with other somatic cells. In each asymmetric stem cell division, the stem lineage may retain the original DNA templates, with all new DNA copies segregating to the transit lineage. If most mutations occur in the production of new DNA strands, then most mutations would segregate to the transit lineage, and the stem lineage would accumulate fewer mutations per cell division (Merok et al. 2002; Potten et al. 2002; Smith 2005; Karpowicz et al. 2005). In addition, stem cells may be particularly prone to apoptosis in response to DNA damage, killing themselves rather than risking repair of damage (Potten 1998; Bach et al. 2000).

Third, compartmental organization of tissues reduces the opportunity for competition and selection between lineages. In the epidermis and intestine, each stem lineage clonally renews a small, well-defined sector of tissue. The whole tissue spans numerous separate, noncompeting cell lineages. The colon has about 10^7 such compartments, called crypts. A mutation in one compartment remains confined to that location, unless the mutation provides an invasive phenotype that causes cells to break into neighboring compartments. Put another way, the compartmental structure reduces competition between cellular lineages by providing a barrier to clonal expansion, thus limiting the number of descendant cells that carry a noninvasive mutation.

To summarize Cairns' view, asymmetric mitoses of stem cells reduce mutation accumulation within lineages, and compartmentalization reduces competition and selection between lineages. Symmetric mitoses and exponential cell lineage expansion increase the risk of cancer progression. I follow up on these issues in a later chapter on cell lineages.

4.7 Hypermutation, Chromosomal Instability, and Selection

Two process may accelerate the accumulation of genomic change. First, changes early in progression may accelerate the production of subsequent changes. Second, competition and selection between cell lineages that harbor various genomic changes would favor clonal expansion of more aggressive lines.

ACCELERATION OF VARIATION BY MUTATORS

Burdette (1955, p. 218) nicely summarized the potential role of mutators in early stages of progression: "A logical corollary to the somatic mutation hypothesis is that [inherited] mutants act as mutators." Those mutators would accelerate the accumulation of subsequent somatic changes in cells. Loeb developed the mutator hypothesis through a series of papers (Loeb et al. 1974; Loeb 1991, 1998; Beckman and Loeb 2005).

Nowell (1976, p. 26) emphasized chromosomal instabilities: "It is possible that one of the earliest changes in tumor cells involves activation of a gene locus which increases the likelihood of subsequent nondisjunction or other mitotic errors." Recent reviews of chromosomal instability can be found in Rajagopalan et al. (2003) and Michor et al. (2004).

Selection between Variants

Cairns (1975, p. 200) noted that an increase in the mutation rate per cell division would speed up progression. However, in epithelial tissues renewed by stem cells, each new mutation would remain confined to a single linear history of descent. Thus, Cairns stated that

> Unless such mutagenic mutations confer some survival advantage, however, they will remain confined to the stem cells in which they arise ... Probably more important, therefore, are mutations that affect the interactions of a cell with its neighbours. Any mutation that gives a stem cell the ability to move out of its compartment in an epithelium may cause it to form an expanding clone of stem cells.

This quote emphasizes the theory of clonal expansion. However, the early theories of clonal expansion focused only on the consequences of expansion. By contrast, Cairns emphasizes the processes that limit competition, and the types of cellular changes that would bypass those limits and promote competition between lineages. Put another way, the early theories focused on the consequences of selection, and the later theories beginning with Cairns emphasized the mechanisms involved in such selection.

The debate continues about the relative importance of mutators versus selection and clonal expansion (Sieber et al. 2005). Tomlinson et al. (1996) reviewed the issues in favor of selection, arguing against the need to invoke mutators in order to explain the incidence of cancer.

4.8 Epigenetics: Methylation and Acetylation

Many theoretical issues have turned on the rate of transition between key stages in progression. I mentioned the concerns that the commonly accepted somatic mutation rate of about 10^{-6} mutations per gene per cell division seemed too low to some investigators to explain how multiple changes could accumulate.

One recurring problem concerns the definition of "mutation" (Burdette 1955). I am interested in kinetics, so I tend to follow those authors who use the term "mutation" rather loosely for heritable genomic changes that influence progression. Other authors, interested in the particular mechanisms of change that underlie progression, emphasize the distinctions between different kinds of genomic changes.

An early distinction arose between point mutations to particular bases and chromosomal instability, which causes a variety of broad karyotypic changes that often affect dosage and gene expression. Some have argued that mutations causing chromosomal instability likely arise early in progression in many tumors (Nowell 1976; Rajagopalan et al. 2003; Michor et al. 2004). Such chromosomal instability could explain the accumulation of numerous genetic changes in a cell lineage, ultimately leading to malignant disease.

Recent evidence points to an important role for various epigenetic changes in contributing to the overall rate of genomic changes in progression. Epigenetic changes include methylation and acetylation of histone proteins and methylation of DNA (Kuo and Allis 1998; Breivik and Gaudernack 1999b; Wang et al. 2001; Jones and Baylin 2002; Egger et al. 2004; Feinberg and Tycko 2004; Fraga et al. 2005; Genereux et al. 2005; Hu et al. 2005; Robertson 2005; Seligson et al. 2005; Sontag et al. 2006). Both methylation and acetylation can strongly influence gene expression, and both tend to be inherited through a cellular lineage. Complex molecular regulatory systems control these epigenetic processes, determining the rate of change and the stability of inherited changes. The regulatory systems are often perturbed in tumors, causing enhanced rates of epigenetic changes—a different mechanistic form of the mutator phenotype.

With regard to kinetics, epigenetic changes simply provide another contributing factor to the speed at which rate-limiting steps in progression may be passed. If one includes epigenetic change, it may not be so hard to explain how cell lineages accumulate multiple hits over the course of a lifetime. With regard to mechanism, some have proposed that epigenetic change presents a new paradigm of progression (Prehn 2005), but my focus remains on kinetic issues.

4.9 Summary

This chapter completes the background on biological observations of incidence and progression, and on the history of theories to explain patterns of incidence. These background chapters discussed quantitative

theories but did not develop any of the quantitative methods or conclusions. To build a stronger quantitative understanding of the causes of cancer, we need to expand the theory and tie the theory more closely to testable predictions about how particular genetic or physiological processes shift incidence. The next chapter begins development of the quantitative theory by providing a gentle introduction to the mathematical models and to why those models can help to understand cancer.

PART II

DYNAMICS

5 Progression Dynamics

Progression depends on various rate processes, such as the rate of somatic mutation and the time for a solid tumor to build a blood supply. To link rate processes to the observed age-onset curves of cancer incidence, one must understand how the processes combine to determine the speed of progression. This chapter introduces the quantitative theory that links carcinogenic process and incidence.

The first section provides background on mathematical theories of progression. The general approach begins with the assumption that cancer develops through a series of stages. This assumption of multistage progression sets the framework in which to build particular models of progression dynamics. Within this framework, I argue in favor of simple theories that make comparative predictions. If one understands how a particular process affects progression, then one should be able to predict how altering that process changes progression dynamics.

The second section lists some of the observations on cancer incidence that a theory should seek to explain. These observations set the target for mathematical theory and emphasize the need to link progression dynamics to incidence.

The third section introduces the classical model of multistage progression. This model predicts an approximately linear relation between incidence and age when plotted on log-log scales. The observed patterns match this prediction for several cancers. However, the fit of observations to theory is not by itself particularly informative. To make further progress, I emphasize the need for comparative theories. I briefly mention one comparative theory that follows from the classical multistage model: the ratio of incidence rates between two groups depends on the difference in the number of rate-limiting steps in progression. I develop that theory in later chapters.

The fourth section discusses why one should bother with abstract theories that often run ahead of empirical understanding. The main reason is that we are not likely to have much luck in understanding real systems if we cannot understand with simple logic how various processes could in principle combine to influence progression. In addition, it helps to have a toolbox of possible explanations that one thoroughly understands. Such understanding prevents the common tendency to latch onto the first available explanation that seems to fit the data, without full consideration of reasonable alternatives.

The fifth section presents the equations for a simple model of progression through a series of stages. I emphasize that the equations are completely equivalent to a simple diagram that illustrates the flow between stages of progression. The equations introduce the notation and structure of a formal model, paving the way for more detailed analysis in the following chapters.

The sixth section develops technical definitions for incidence and acceleration that follow from the formal specification of the model in the previous section. Incidence provides the key measure of occurrence for cancer: the cases of cancer per year, at each age, for a given population of individuals. Incidence is a rate—cases per year—just as velocity is a rate. Acceleration is the rate of change in incidence with age: how fast incidence increases or decreases as individuals become older. Theories about the carcinogenic role of particular biochemical mechanisms must ultimately link those mechanisms to their effects on incidence and acceleration.

5.1 Background

MULTISTAGE PROGRESSION IS A FRAMEWORK, NOT A HYPOTHESIS

Most mathematical models of cancer progression descend from Armitage and Doll's (1954) paper on multistage theory. The phrase "multistage theory" has led to some confusion. A multistage model simply assumes that cancer does not arise in a single step—an assumption supported by much evidence. So, "multistage theory" is not really a particular theory; it is a framework that describes the kind of dynamical processes used to model progression through multiple stages.

This framework provides tools to develop testable quantitative hypotheses that link progression dynamics to the curves of age-specific cancer incidence. Progression dynamics also provides a notion of causality: a process causes cancer to the extent that the process alters the age-specific incidence curve.

The Importance of Comparative Hypotheses

A mathematical analysis for the age of cancer onset depends on several parameters. Those parameters might include the number of stages in progression, the somatic mutation rate that moves a tissue from one stage to the next, the number of cells in the tissue, and the precancerous rate of cell division. Given values for those parameters, the mathematical model generates an age-specific incidence curve.

A mathematical model may be used in two different ways: fit or comparison.

A fit chooses values for all parameters that minimize the distance between the predicted and observed age-specific incidence curves. A good fit provides a close match between prediction and observation. A good fit also uses realistic values for parameters such as rates of mutation and cell division.

A comparison sets an explicit hypothesis: as a parameter changes, the model predicts a particular direction of change for the age-specific incidence curve. For example, an inherited mutation may reduce by one the number of stages that must be passed during progression. Mathematical models predict that fewer stages cause the incidence curve to have a lower slope and to shift to earlier ages (higher intercept). I will show data that support this comparative prediction.

FITTING

One can fit theory to observation, but the match usually arises because a model with several parameters creates a flexible manifold that conforms to the data. Even when one constrains parameter estimates to realistic values, an incorrect model with several parameters often has great flexibility to conform to the shape of the data. A fit is achieved so easily that such a model, fitting widely and well, actually explains very little. As Dyson (2004) tells it:

> In desperation I asked Fermi whether he was not impressed by the agreement between our calculated numbers and his measured numbers. He replied, "How many arbitrary parameters did you use for your calculations?" I thought for a moment about our cutoff procedures and said, "Four." He said, "I remember my friend Johnny von Neumann used to say, with four parameters I can fit an elephant, and with five I can make him wiggle his trunk." With that, the conversation was over.

Several mathematical methods test the quality of a fit. But technical fixes do not overcome the main difficulty: mathematical models fail to capture the full complexity of multidimensional problems such as cancer. If a model does become sufficiently complex, one has so many parameters that fitting almost anything is accomplished too easily.

Although a good fit means little, a lack of fit also provides little insight: lack of fit means only that one does not have exactly the right model. However, one rarely has exactly the right model. So, by lack of fit, one may end up rejecting a theory that in fact captures much of the essential nature of a process but misses one aspect.

Finally, another common approach considers the realism of parameter estimates obtained from the data. For example, when fitting a model, how close do the estimated mutation rates match values thought to be realistic? However, parameter estimates can only be compared to realistic values when one has a complete model. In incomplete models, the parameter estimates change to make up for processes not included in the model. So the realism of parameter estimates provides a test only when fitting a complete model that captures the full complexity of a process. But for cancer and for most interesting biological phenomena, we do not have complete models and probably never will have complete models.

Models do have great value in spite of the difficulties of drawing conclusions by fitting to the data. The key is to develop and test theories in a comparative way.

COMPARISON

A comparison is simple to formulate, understand, and test. Consider the following prediction: as the number of steps in progression declines, the slope of the incidence curve decreases. To test this, one has to measure a relative change in the number of steps and a relative change in the slope of the incidence curve. This test can be accomplished by comparing the incidence curves between genotypes, where one genotype has a mutation that abrogates a suspected rate-limiting step in progression.

A comparative prediction allows tests of causal hypotheses. If I understand what causes cancer, then I can predict how incidence curves change as I change the underlying parameters of cancer dynamics.

The limited role of mathematics and quantitative studies in much of biology follows from a fatal attraction to fitting complex models. Simple comparative models are often rejected a priori because they do not contain all known processes. The reasoning seems to be: how can a model be useful if a known process is left out? All known processes are added in; fits are obtained; little is learned; quantitative analysis is abandoned.

A model is not a synthesis of all known observations; a model is a tool to test one's ability to predict the behavior of a system. If one cannot say how the system changes when perturbed, then one does not understand the system. To study perturbations most effectively, formulate and test the simplest comparative theories.

5.2 Observations to Be Explained

In this section, I briefly list a few puzzles—just enough to set the context. Chapter 2 provided a more complete review of the observations on age-specific incidence.

The difference in incidence curves between inherited and sporadic cancers provides the most striking observation (Knudson 1971, 2001). In the simplest case, the inherited form of a cancer arises in those who carry a defect in a single allele. For example, a carrier with a mutant *APC* allele typically develops numerous independent colon tumors in midlife. By contrast, sporadic (noninherited) cases mostly occur later in life.

The comparison between inherited and sporadic incidence curves presents an opportunity to test how particular mutations affect the rate of cancer progression. Figure 2.6 compares incidence data between sporadic cancers and inherited cancers in carriers of a mutation to a single allele. Comparison of incidence curves between experimentally controlled genotypes of rodents provides an exceptional opportunity to test hypotheses. Figure 2.7 illustrates the sort of data that can be obtained. Later, I will provide methods to analyze those data with regard to quantitative models of progression dynamics.

Six additional patterns in the incidence data suggest the kinds of puzzles that dynamical theories of progression must explain.

First, incidence accelerates slowly with age for some cancers, such as melanoma, thyroid, and cervical cancers. By contrast, other cancers accelerate more rapidly with age, such as colorectal, bladder, and pancreatic cancers (Figure 2.3).

Second, the acceleration of cancer incidence with age declines at later ages for the common epithelial cancers—breast, prostate, lung, and colorectal (Figure 2.3). Several other cancers also show a steady and sometimes rather sharp decline in acceleration at later ages. In some cases, the patterns of acceleration differ between countries (see Appendix). On the whole, declines in acceleration later in life appear to be typical for many cancers.

Third, several cancers show very high early or midlife accelerations, sometimes with accelerations at early ages rising to a midlife peak (Figure 2.3). For example, prostate cancer has an exceptionally high midlife peak (Figure A.2); leukemia (Figure A.6) and in some cases colon cancer (Figure A.4) show rises in early life.

Fourth, smokers who quit by age 50 have a lower acceleration in lung cancer risk later in life than do those who never smoked or who continue to smoke (Figure 2.8).

Fifth, exposure to a carcinogen often causes the median number of years to tumor formation to decline linearly with dosage when measured on log-log scales (Figures 2.10, 2.11).

Sixth, given a set of individuals who have suffered breast cancer at a particular age, the close relatives of those individuals have high and nearly constant annual risk (zero acceleration) for breast cancer after the age at which the affected individuals were diagnosed. By contrast, individuals whose relatives have not suffered breast cancer have lower risk per year, but their risk accelerates with age (Peto and Mack 2000).

These observations provide a sample of interesting puzzles, most of which have yet to be explained in a convincing way. Dynamical models of cancer progression provide the only source of plausible hypotheses to explain the range of observed patterns.

5.3 Progression Dynamics through Multiple Stages

Models of progression dynamics analyze transitions through stages. The simplest type of model follows progression through a linear sequence. This linear model arose over 50 years ago, when people first observed clear patterns in the age-specific incidence of cancer.

Figure 5.1 illustrates the type of pattern that was apparent to early observers: the incidence of colorectal cancer increases in a roughly linear

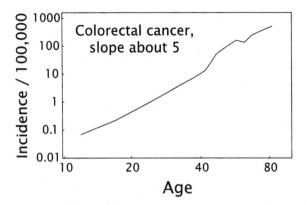

Figure 5.1   Age-specific incidence for colorectal cancer. Data for all males from the SEER database (www.seer.cancer.gov) using the nine SEER registries, year of diagnosis 1992–2000.

way with age when plotted on log-log scales. In an earlier chapter, Figure 2.2 showed that log-log plots of incidence are approximately linear for many cancers.

The line in Figure 5.1 fits a model in which

$$I = ct^{n-1},$$

where I is cancer incidence at age t, the exponent $n - 1$ determines the rate of increase in cancer incidence with age, and c is a constant. Taking the logarithm of both sides of this equation gives the log-log scaling shown in the figure

$$\log (I) = \log (c) + (n - 1) \log (t),$$

in particular, the figure plots $\log(I)$ versus $\log(t)$. The line in Figure 5.1 has a slope of $n - 1 \approx 5$.

The linear rise on log-log scales means that incidence is increasing exponentially with age in proportion to t^{n-1}. In the early 1950s, several authors wondered what might explain this exponential rise in incidence with age (Frank 2004c; Moolgavkar 2004).

Fisher and Hollomon (1951) recognized that cancer incidence would increase as t^{n-1} if transformation required n independent steps. The argument is roughly as follows. Suppose each step happens at a rate of

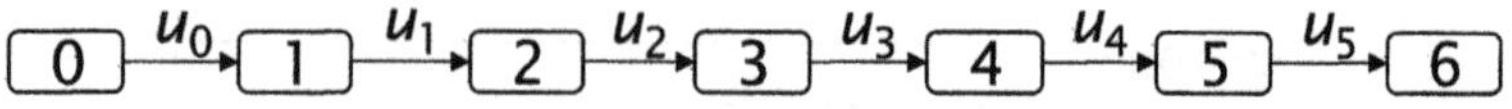

Figure 5.2 Multistage model of cancer progression. Individuals are born in stage 0. They progress from stage 0 through the first transition to stage 1 at a rate u_0, then to stage 2 at a rate u_1, and so on. Severe cancer only arises after transition to the final stage. With regard to epidemiology, the rate at which individuals enter the final stage, $n = 6$ in this case, is approximately proportional to t^{n-1} as long as cancer remains rare and the u_i's are not too different from each other.

u per year, where u is a small rate. The probability of any step having happened after t years is $1 - e^{-ut} \approx ut$. At age t, the probability that $n - 1$ of the steps has occurred is approximately $(ut)^{n-1}$, and the rate at which the final step happens is u, so the approximate rate (incidence) of occurrence at time t is proportional to $u^n t^{n-1}$.

Nordling (1953) and Armitage and Doll (1954) emphasized that the different steps may happen sequentially. There are $n - 1!$ different orders in which the first $n - 1$ steps may occur. If we assume they must occur in a particular order, then we divide the incidence calculated in the previous paragraph, $u^n t^{n-1}$, by $n - 1!$ to obtain the approximate value for passing n steps at age t as

$$I_n(t) \approx \frac{u^n t^{n-1}}{n - 1!}. \tag{5.1}$$

Armitage and Doll (1954) developed this theory of sequential stages for the dynamics of progression—the multistage theory of carcinogenesis as illustrated in Figure 5.2.

This basic model provides a comparative prediction for the relative incidence of sporadic and inherited cancers (Frank 2005). Suppose that normal individuals develop sporadic cancer in a particular tissue after n steps. Individuals carrying a mutation develop inherited cancer after $n - 1$ steps, having passed one step at conception by the mutation that they carry. Using Eq. (5.1) for n steps versus $n - 1$ steps, the incidence ratio of sporadic to inherited cancers at any age t is

$$R = \frac{I_n}{I_{n-1}} \approx \frac{ut}{n - 1}.$$

In Chapter 8, I will develop this comparative prediction and apply it to data from retinoblastoma and colon cancer. That application will show how a simple comparative theory can link the genetics of cancer progression to the age of cancer incidence.

5.4 Why Study Quantitative Theories?

An ordered, linear sequence leaves out many of the complexities of carcinogenesis. However, it pays to begin with this simple model, to understand all of its logical consequences, and to study how well that model can predict changes in incidence. Following on the simple model, we can begin to explore alternatives, such as parallel lines of progression in different cellular lineages or incidence aggregated over different pathways.

After I have analyzed the basic model, I will explore a range of more complex assumptions, because we need to understand the possible alternative explanations for observed patterns. Without broad conceptual understanding, there is a tendency to latch onto the first available explanation that fits the data without full consideration of reasonable alternatives. The theory I develop will run ahead of empirical understanding, but if used properly, this is exactly what theory must do.

Another issue concerns the definition of stages and rate-limiting steps. To address this issue, we must consider what we wish to accomplish with mathematical models. The models are tools, so we need be concerned only about defining stages and rate-limiting steps in ways that help us to achieve particular goals for particular problems.

Sometimes we may formulate a model in a very abstract, nonbiological way, for example, to study how variation in rates of transition between stages influences age-onset patterns. In this case, stages remain abstract notions that we manipulate in a mathematical model in order to understand the logical consequences of various assumptions. In other cases, we may try to match the definition of stages and rates to the biological details of a particular cancer. A stage may, for example, be an adenoma of a particular size, histology, and genetic makeup. A transition between stages may occur at the rate of a somatic mutation to a particular gene.

5.5 The Basic Model

Assume that cancer progression requires passage through n rate-limiting steps, each step moving through the sequence of tumor progression to the next stage. A step could, for example, be mutation to *APC* or *p53*, as in colorectal cancer progression. But for now, I just assume that such steps must be passed.

Not all changes during tumor development limit the rate of progression. A necessary change may happen very quickly following, for example, expansion of a precancerous tumor to a large size. Such a step is necessary for progression but does not limit the rate of progress, and so does not determine the ages at which individuals carry tumors of particular stages. I develop the basic theory under the assumption that whatever determines a rate-limiting step, tumor progression requires passing n such steps to develop into cancer. This section follows the derivations given in Frank (2004a).

I gave a picture of the basic model in Figure 5.2. That picture formally describes a set of differential equations. Because the picture and the equations present the same information, one may choose to focus on either. The equations are

$$\dot{x}_0(t) = -u_0 x_0(t) \tag{5.2a}$$

$$\dot{x}_j(t) = u_{j-1}x_{j-1}(t) - u_j x_j(t) \qquad i = j, \ldots, n-1 \tag{5.2b}$$

$$\dot{x}_n(t) = u_{n-1}x_{n-1}(t), \tag{5.2c}$$

where $x_i(t)$ is the fraction of the initial population born at time $t = 0$ that is in stage i at time t, with time measured in years. Usually, I assume that when the cohort is born at $t = 0$, all individuals are in stage 0, that is, $x_0(0) = 1$, and the fraction of individuals in other stages is zero. As time passes, some individuals move into later stages. The rate of transition from stage i to stage $i + 1$ is u_i. The $\dot{x}$'s are the derivatives of x with respect to t.

5.6 Technical Definitions of Incidence and Acceleration

Two ways to characterize age-onset patterns play an important role in analyzing cancer data and studying theories of cancer progression. Incidence is the rate at which individuals develop cancer at particular ages. Acceleration is the change in incidence rates. For example, positive acceleration means that incidence increases with age.

This section provides some technical details for the definitions of incidence and acceleration. One can get a rough idea of the main results without these details, so some readers may wish to skip this section and come back to it later.

Individuals who move into the final, nth stage develop cancer. They pass into the final stage at the age-specific incidence rate $\dot{x}_n(t)$, which is roughly the probability of developing cancer per year at age t. The age-specific incidence is the fraction of all individuals in the cohort who develop cancer for the first time at age t, which is the probability of developing cancer at age t divided by the fraction of individuals, $S(t)$, who have not yet developed cancer by that age. In symbols, we write that the age-specific incidence is $I(t) = \dot{x}_n(t)/S(t)$.

The incidence, $I(t)$, is the rate at which cancer cases accumulate at a particular age. I frequently refer to the acceleration of cancer, which is how fast the rate, $I(t)$, changes at a particular age, t. The most useful measure of acceleration in multistage models scales incidence and time logarithmically (Frank 2004a, 2004b).

Use of logarithms provides a scale-free measure of change. In other words, differences on a logarithmic scale summarize percentage change in a variable independently of the value of the variable. This can be seen by examining the derivative of the logarithm for a variable x, which is

$$d \log (x) = \frac{dx}{x}.$$

The right side is the change in x divided by x, which measures the fractional change in x independently of how large or small x is.

For example, if we wanted to measure the percentage increase in the age-specific incidence for a given percentage increase in age, then we need to measure in a scale-free way changes in both age-specific incidence and age. We obtain a scale-free measure by defining the log-log acceleration (LLA) at age t as

$$\text{LLA}(t) = \frac{dI(t)/I(t)}{dt/t} = \frac{d \log(I(t))}{d \log(t)}. \tag{5.3}$$

The derivative of incidence, $dI(t)/dt$, is the age-specific acceleration, so LLA is just a normalized (nondimensional) measure of age-specific acceleration.

5.7 Summary

This chapter introduced the quantitative tools needed to build models of cancer progression. Such models make predictions about how particular genetic or physiological changes alter age-specific incidence. The ability to make such predictions successfully defines a causal understanding of cancer. The next chapter begins my mathematical analysis of the ways in which particular causes affect age-specific incidence.

6 Theory I

To test hypotheses about how particular biochemical processes affect cancer, we need quantitative predictions for how biochemical changes alter the age of cancer onset. This chapter develops the quantitative theory of progression dynamics.

The first section outlines my strategy for presentation. I divide each quantitative analysis into a précis that gives the main points, a mathematical presentation of the analytical details, and a set of conclusions.

The second section solves the basic model of multistage progression dynamics. In that model, individuals progress through a series of stages with the same constant transition rate from each stage to the next. That model follows the classical analysis of multistage progression, leading to the conclusion that a log-log plot of cancer incidence versus age is approximately linear with a slope of $n - 1$, where n is the number of rate-limiting steps in progression. The slope of $n - 1$ measures the acceleration of cancer with age. I present an exact solution for the model, which shows that, under some conditions, the incidence curve flattens late in life and drops below the linear approximation, causing a late-life decline in acceleration.

The third section analyzes parallel lines of progression within individuals. The models follow the stages of cells or tissue compartments, in which different cells or compartments may be in different stages of progression within the same tissue. The greater the number of independent lines of progression, the slower progression must be in each line to keep the overall incidence from rising to very high levels. The smaller the number of lines, the more strongly acceleration tends to decline later in life.

The fourth section discusses how incidence changes when the rates of transition vary between different stages in progression. The greater the variation in rates of transition, the more strongly the acceleration of cancer tends to decline with advancing age.

The fifth section studies what happens when rates of transition vary with age. Rates may increase with age if DNA repair capacity or other checks on cell-cycle integrity decline with age. Alternatively, rates of transition may rise when a precancerous cell expands into a large clone, in which a subsequent change to any one of the clonal cells could cause progression to the next stage in carcinogenesis. As the clone grows larger, the target size for a transition increases. Time-varying rates often cause a rise in acceleration to a midlife peak, followed by a late-life decline in acceleration.

6.1 Approach

This chapter and the following one develop the theory of progression dynamics. Most of the sections contain some mathematics. I use the following structure to make the presentation accessible. A section with mathematics begins with a précis that highlights the main results. The mathematical details follow, often with some illustrations to emphasize the key points. The section ends with a brief statement of the conclusions.

I developed much of the following original theory for this book. Although the overall structure and many of the particular results are new, my mathematical work grew from a rich and highly developed field. I gave an overview of the history in Chapter 4. I particularly wish to acknowledge the pioneering contributions of Armitage and Doll, Knudson, and Moolgavkar, who have been most influential in my own studies.

6.2 Solution with Equal Transition Rates

Précis

I start with the linear chain of stepwise progression illustrated in Figure 5.2. No type of cancer will always follow the same steps with fixed transition rates between steps. But a thorough understanding of the simplest case puts us in a better position to study more realistic assumptions.

In this section, I assume that the transitions between steps happen at the same rate, u, and that everyone is born in stage 0. Individuals who progress through the nth stage develop cancer.

With these assumptions, the fraction of the population at age t in each precancerous stage is given by the Poisson distribution with a mean of ut. Intuitively, ut would be the average number of transitions passed if there were unlimited stages, because u is the transition rate per stage and t is the time that has elapsed. So the probability of i transitions among the precancerous stages follows the standard Poisson process.

If cancer remains uncommon by age t, then incidence is $I(t) \approx kt^{n-1}$, where $k = u^n/(n-1)!$. On log-log scales,

$$\log(I(t)) \approx \log(k) + (n-1)\log(t).$$

The log-log acceleration is

$$\text{LLA}(t) \approx n - 1.$$

This is the classical result that log-log plots of incidence versus age will be approximately linear with a slope of $n - 1$ (Armitage and Doll 1954).

When a significant fraction of individuals develops cancer, the log-log incidence plot tends to accelerate more slowly at later ages, causing the curve to flatten late in life and drop below the linear approximation. The following details provide an exact solution for this simple model. The exact solution shows how acceleration declines with age.

Details

I introduced the basic model in Eqs. (5.2) of the previous chapter. I repeat those equations here to provide the starting point for further analysis

$$\dot{x}_0(t) = -u_0 x_0(t) \qquad\qquad\qquad\qquad\qquad\qquad (6.1a)$$

$$\dot{x}_j(t) = u_{j-1} x_{j-1}(t) - u_j x_j(t) \qquad i = j, \ldots, n-1 \qquad (6.1b)$$

$$\dot{x}_n(t) = u_{n-1} x_{n-1}(t), \qquad\qquad\qquad\qquad\qquad (6.1c)$$

where $x_i(t)$ is the fraction of the initial population born at time $t = 0$ that is in stage i at time t, with time measured in years. Usually, I assume that when the cohort is born at $t = 0$, all individuals are in stage 0, that is, $x_0(0) = 1$, and the fraction of individuals in other stages is zero. As time passes, some individuals move into later stages. The rate of transition from stage i to stage $i + 1$ is u_i. The $\dot{x}$'s are the derivatives of x with respect to t.

If the transition rates are constant and equal, $u_j = u$ for all j, then we can obtain an explicit solution for the multistage model (Frank 2004a). This provides a special case that helps to interpret more complex assumptions that must be evaluated numerically. The solution is $x_i(t) =$

$e^{-ut}(ut)^i/i!$ for $i = 0, \ldots, n-1$, with the initial condition that $x_0(0) = 1$ and $x_i(0) = 0$ for $i > 0$. Note that the $x_i(t)$ follow the Poisson distribution for the probability of observing i events when the expected number of events is ut.

In the multistage model above, the derivative of $x_n(t)$ is given by $\dot{x}_n(t) = ux_{n-1}(t)$. From the solution for $x_{n-1}(t)$, we have $\dot{x}_n(t) = ue^{-ut}(ut)^{n-1}/n-1!$. Age-specific incidence is

$$I(t) = \frac{\dot{x}_n(t)}{1 - x_n(t)} = \frac{\dot{x}_n(t)}{\sum_{i=0}^{n-1} x_i(t)} = \frac{u(ut)^{n-1}/n-1!}{\sum_{i=0}^{n-1}(ut)^i/i!}, \qquad (6.2)$$

and log-log acceleration from Eq. (5.3) is

$$\text{LLA}(t) = \frac{dI(t)/I(t)}{dt/t} = n - 1 - ut(S_{n-2}/S_{n-1}), \qquad (6.3)$$

where $S_k = \sum_{i=0}^{k}(ut)^i/i!$.

The total fraction of the population that has suffered cancer by age t—the cumulative probability—is

$$x_n(t) = 1 - e^{-ut}S_{n-1}. \qquad (6.4)$$

This analysis does not explicitly follow causes of mortality other than cancer. Frank (2004a) analyzed the case in which each stage has a constant transition rate to the next stage, u, as above, and also a constant mortality rate from other causes, d. With constant mortality, d, the only change in the solution arises in the expression $x_i(t) = e^{-(u+d)t}(ut)^i/i!$ for $i = 0, \ldots, n-1$, in particular, with extrinsic mortality, we must use $e^{-(u+d)t}$ in the solution rather than e^{-ut}. Because these exponential terms arise in both the numerator and denominator of the expression for incidence and so cancel out, extrinsic mortality does not affect the incidence and acceleration solutions given here. The classes x_i for $i = 0, \ldots, n-1$ can be interpreted as those individuals alive and tumorless at different stages in progression.

CONCLUSIONS

This simple model shows the tendency of incidence to increase with age in an approximately linear way on log-log scales. The increase in incidence with age occurs because individuals progress through multiple precancerous stages. Many processes cause departures from log-log linearity. The following sections explore some of the ways in which progression affects the shape of the age-incidence curve.

6.3 Parallel Evolution within Each Individual

The model in the previous section assigns each individual in the population to a particular stage of progression. Sometimes, it may make more sense to consider the stage of particular cells or tissue compartments within a single individual. Different components may be in different stages of progression.

I described in Chapter 3 how colorectal cancer initiates in individual crypts, perhaps with mutations that occur to a particular stem cell within a crypt. So we might choose to focus on different stages of progression in different crypts or stages of progression in different stem cell lineages. The human colon has about 10^7 crypts, and a slightly higher number of stem cell lineages, so each individual has many parallel, independent lines of progression.

Précis

Suppose each individual has L independent lines of progression. We start by calculating the rate of transition into the final, cancerous stage for each independent line—the incidence per line. The incidence per individual is the rate at which one of the L lines moves into the final, cancerous state. The incidence per individual is simply L multiplied by the incidence per line: the cancer rate rises linearly with the number of independent lines that can fail.

If we fix the rate of progression per line, then the number of independent lines does not affect log-log acceleration. However, if we wish to keep constant the overall probability per individual of developing cancer by a certain age, then as the number of lines increases, the probability of cancer per line must decline. Interestingly, slower per-line transformation keeps acceleration higher through later ages, because slow transformation maintains a high number of stages remaining in progression.

Details

Let the number of parallel lines of evolution within each individual be L. We now have to consider progression hierarchically. Within each individual, cancer arises as soon as one of the L lines progresses to the nth stage. For each independent line, the probability of progressing to the final malignant state by time t is $x_n(t)$. The cumulative probability of cancer is the probability that at least one of the L lines has progressed to the malignant state. This cumulative probability of cancer by age t is

$$p(t) = 1 - [1 - x_n(t)]^L. \tag{6.5}$$

For large L and small $x_n(t)$, the Poisson approximation is very accurate, $p(t) \approx 1 - e^{-x_n(t)L}$. The Poisson distribution with mean $x_n(t)L$ gives the distribution of the number of independent tumors per individual at age t.

Incidence is the rate of new cases divided by the fraction of the population at risk. Using the definition for $p(t)$ in Eq. (6.5) and dropping t from the notation,

$$I = \frac{\dot{p}}{1-p} = \frac{L\dot{x}_n (1 - x_n)^{L-1}}{(1 - x_n)^L} = \frac{L\dot{x}_n}{1 - x_n}.$$

Comparing this result with Eq. (6.2) shows that having L independent lines of progression within an individual simply increases incidence by a constant value L. Log-log acceleration is independent of constant multiples of incidence, as shown in Eq. (5.3), so log-log acceleration is independent of L and is given by Eq. (6.3).

What does change is the value of u that one must assume in order for a certain total fraction of the population to have cancer by a particular age, T. If the fraction of the population with cancer is m, then u is obtained by solving $m = p(T)$ for u, using Eq. (6.5) for $p(T)$ and Eq. (6.4) for $x_n(T)$. As the number of independent lines, L, increases, slower transitions must be assumed to give the same overall incidence. This reduction of u causes each line to progress more slowly, but, by chance, one of the many separate lines within an individual progresses to the final stage with probability $p(T)$.

If, under the assumptions of this model, individuals rarely have more than one independent tumor, then the per-line probability of progression is approximately m/L, the total probability of progression per individual, m, divided by the number of lines, L. It is often most informative to evaluate progression on a per-line basis and to present results for particular levels of $m/L \approx x_n(T)$. In this model, multiple tumors per individual are rare when $m = p(T) < 0.2$.

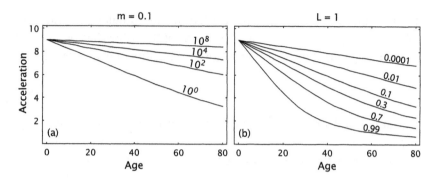

Figure 6.1   Acceleration of cancer incidence in a multistage model calculated from Eq. (6.3). For all curves: $n = 10$; the cumulative probability of cancer by age $T = 80$ is $m = p(80)$; and L is the number of independent lines of progression within each individual. (a) The cumulative probability of cancer by age 80 is set to $m = 0.1$. The values on each curve show L. The values of u were obtained by solving $m = p(80)$, yielding for the curves from top to bottom: 0.00757, 0.0209, 0.0373, 0.0778. (b) The number of independent lines is set to $L = 1$. The values on each curve show m. The values of u were obtained by solving $m = p(80)$ in Eq. (6.5), yielding for the curves from top to bottom: 0.0275, 0.0516, 0.0778, 0.1017, 0.1423, and 0.2348. The two panels show results for separately varying values of m and L, but for $m < 0.2$, each curve depends only on the ratio m/L.

Conclusions

Figure 6.1 shows how acceleration declines with age in multistage progression. The decline in acceleration occurs because individuals pass through the early stages of progression as they age. In this model, all lines in all individuals are in stage 0 at birth, with n steps remaining. Acceleration at birth is $n - 1$, as shown in the figure. Suppose at a later age that all lines have progressed through a steps. Then at that age they have $n - a$ steps remaining, and an acceleration of $n - a - 1$ (Figure 6.2).

In reality, all lines do not progress equally with age. The different lines in separate individuals move stochastically through the various stages of transformation. At any particular age, there is a regular probability distribution of tissue components that have progressed to particular precancerous stages or all the way to the final, malignant stage.

The acceleration at any age depends on the distribution of individual tissue components into different stages of progression (Figure 6.3). For this simple model, acceleration at a particular age is approximately $n -$

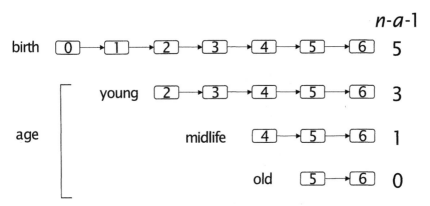

Figure 6.2   Cause of declining acceleration with age in multistage progression. The top line shows the six stages that a newborn must pass through in this case. As individuals grow older, many may pass through the early stages. This example shows rapid progression to emphasize the process. Here, most individuals have passed to stage 2 by early life, so the acceleration at this age, the number of steps remaining minus one, is three. By midlife, two steps remain, causing an acceleration of one. By late life, all individuals who have not developed cancer have progressed to the penultimate stage, and so with one stage remaining, they have an acceleration of zero. Redrawn from Frank (2004d).

$\bar{a} - 1$, where $\bar{a}$ is the average stage of progression among those lines that have not progressed to the nth stage.

6.4 Unequal Transition Rates

When there are many independent lines in a tissue, then the probability that any particular line progresses to cancer must be low. For example, in the colon, L is probably between 10^7 and 10^8, because there are about 10^7 independent tissue compartments (crypts). If the lifetime incidence is about $m = 10^{-1}$, then the incidence per line is approximately m/L, which is small.

When the progression per line, m/L, is small, as in the upper curves of Figure 6.1a, and the transition rates between steps are equal, then acceleration declines relatively little with age. Stable acceleration occurs because most lines remain in the early stages even among older individuals (Figure 6.3, upper panels).

If transition rates differ between stages, then acceleration does decline with age even when the progression per line is small. The top curve

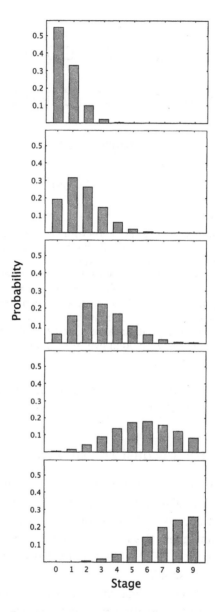

Figure 6.3 Distribution of independent lines of progression across various stages, which depends only on n, u, and t. Here, $n = 10$ and $t = 80$. The stage $n = 10$ is excluded; that stage causes cancer, and the distributions here show the stages among individuals who have not had cancer. The panels from top to bottom correspond to the parameters for the four curves from top to bottom in Figure 6.1a, plus a fifth value of $u = 0.1209$, corresponding to $m = 0.5$ and $L = 1$, for the distribution in the bottom panel of this figure.

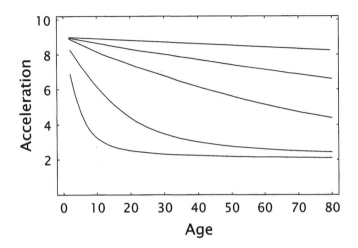

Figure 6.4 Increasing variation in rates of transition reduces acceleration. In this example, there are $n = 10$ steps. Three steps have relatively slow transition rates, $u_0 = u_3 = u_7 = s$, and the other seven steps have fast rates, f. The lifetime risk per line, m/L, was set to 10^{-8} for all curves, so if $L = 10^7$, then the lifetime risk per individual is 0.1. The slow and fast rates are calculated by $s = u^*/d^2$ and $f = u^*d$. For the curves, from top to bottom, $u^* = 0.00962, 0.00963, 0.0119, 0.0238, 0.0516$, and $d = 1, 5, 10, 20, \sqrt{100}$. In all cases, the ratio of fast to slow rates is $f/s = d^3$; the lower the curve, the greater the variation in rates.

in Figure 6.4 shows the nearly constant acceleration with age when transition rates do not differ and $m/L = 10^{-8}$. As the variation in transition rates rises, the curves in Figure 6.4 drop to lower accelerations. (I numerically evaluated Eqs. (6.1) for all calculations in this section.)

Figure 6.5 shows the distribution of lines in different stages at age 80, where the panels from top to bottom match the increasing variation in rates for the curves from top to bottom in Figure 6.4.

Why does rate variation cause a drop in acceleration with age? At birth, all individuals are in stage 0, and there are $n = 10$ steps to pass to get to the final cancerous stage of progression. So, the acceleration is $n - 1 = 9$, independently of the variation in rates, because each of the n steps remains a barrier.

The bottom panel of Figure 6.5 shows the consequences of high variation in rates for the distribution of lines into stages at age 80. The probability peaks for stages 0, 3, and 7 arise because transitions out of those stages are relatively slow compared to all other transitions. The fast transitions between, for example, stages 1 and 2, and between

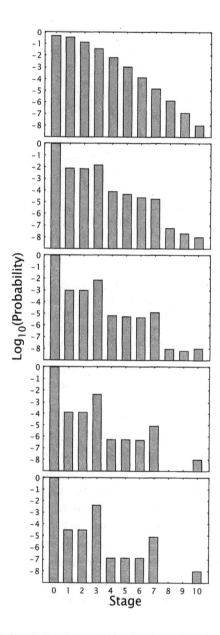

Figure 6.5   Probability that a line will be in a particular stage at age 80. Parameters for the panels here from top to bottom match the curves from top to bottom in Figure 6.4. The expected number of lines in each stage is $p_i L$, where p_i is the probability that a line is in the ith stage, and L is the number of lines. If the number of lines in an individual tissue is $L = 10^7$, then, on a logarithmic scale, the expected number of lines in each stage is $\log_{10}(p_i L) = \log_{10}(p_i) + 7$.

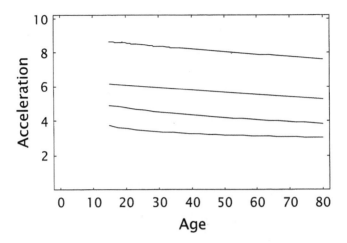

Figure 6.6 Increasing variation in rates of transition reduces acceleration. In this example, there are $n = 10$ steps. The first and last steps are the slowest; the middle steps are the fastest. In particular, $u_i = u_{n-1-i} = u^* k^i$ for $i = 0, \ldots, 4$, with u values chosen so that $m/L = 10^{-8}$. Larger values of k cause greater variation in rates. Greater rate variation reduces acceleration by concentrating the limiting transitions onto fewer steps. Here, for the curves from top to bottom, the values are $k = 2$ and $u^* = 2.245 \times 10^{-3}, 2.715 \times 10^{-4}, 6.85 \times 10^{-5}, 2.66 \times 10^{-5}$. The values of accelerations for ages less than 15 were erratic because of the numerical calculations. At $t = 0$ the acceleration is $n - 1 = 9$.

stages 2 and 3, happen relatively quickly and do not limit the flow into the final, cancerous stage. Only the $n_s = 3$ slow transition rates limit progression, and so acceleration declines to $n_s - 1 = 2$, as shown in Figure 6.4.

In the long run, the slowest steps determine acceleration (Moolgavkar et al. 1999). But the long run may be thousands of years, so we need to consider how acceleration changes over the course of a typical life when rates vary. Figure 6.6 shows a different pattern of unequal rates. In that figure, the first and last transitions happen at the slowest rate, and the rates rise toward the middle transitions. As one follows the curves from top to bottom, the variation in rates increases and the accelerations decline. Figure 6.7 shows the distribution of lines into stages at age 80, with the panels from top to bottom matching the curves from top to bottom in Figure 6.6.

Armitage (1953) presented the classical approximation for unequal rates. However, Moolgavkar (1978) and Pierce and Vaeth (2003) noted

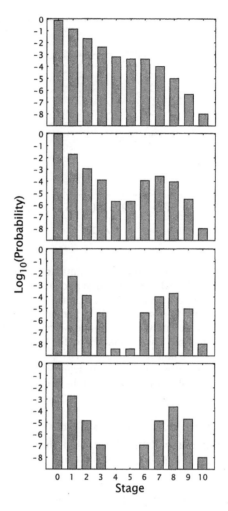

Figure 6.7   Probability that a line will be in a particular stage at age 80. Parameters for the panels here from top to bottom match the curves from top to bottom in Figure 6.6. Probability shown on a $\log_{10}$ scale. If the number of lines in an individual tissue is $L = 10^7$, then, on a logarithmic scale, the expected number of lines in each stage is $\log_{10}(p_i L) = \log_{10}(p_i) + 7$.

that Armitage's approximation can be off by a significant amount. I have avoided using such approximations here and in other sections. With modern computational tools, it is just as easy to obtain exact results by direct calculation of the dynamical system, as I do throughout this book.

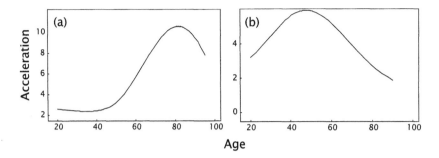

Figure 6.8   Acceleration when all transition rates increase with age. (a) The parameters are $n = 4$, $u = 0.02$, $F = 20$, $a = 8.5$, $b = 1.5$, $T = 100$. (b) The parameters are $n = 4$, $u = 0.012$, $F = 5$, $a = 5$, $b = 5$, $T = 100$.

In summary, unequal rates cause a decrease in acceleration. When there are n_s relatively slow rates, and all other rates are relatively fast, then acceleration early in life starts at $n - 1$ and then declines to $n_s - 1$. When rate variation follows a more complex pattern, increasing variation will usually cause a decline in acceleration, but the particular pattern will depend on the details.

6.5 Time-Varying Transition Rates

Précis

The previous models assumed that transition rates between stages remain constant over time. Many process may alter transitions rates with age. In this section, I analyze two factors that may increase the transition rate between particular stages. In the first model, advancing age may be associated with an increase in transition rates between stages, for example, by an increase in somatic mutation rates (Frank 2004a). In the second model, a cell arriving in a particular stage may initiate a clone of aberrant, precancerous cells. Clonal expansion increases the number of cells at risk for acquiring another change, increasing the rate of transition to the next stage of progression (Armitage and Doll 1957).

Transition rates that increase over time cause a rise in incidence with age, increasing acceleration. The faster transitions also move more older individuals into later stages, causing a late-life decline in acceleration. Thus, increasing transition rates often cause acceleration to rise to a midlife peak, followed by decline late in life (Figures 6.8, 6.9).

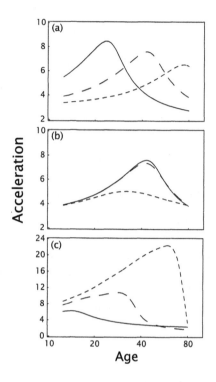

Figure 6.9   Clonal expansion influences patterns of acceleration. (a) Slower clonal expansion shifts peak acceleration to later ages. Parameters for all curves are $n = 4$, $K_i = 1$ for $i = 0, \ldots, n-2$, and $K_{n-1} = 10^6$. The curves have values of $r_{n-1} = 0.4, 0.2, 0.1$ for the solid, long-dash, and short-dash curves, respectively. The mutation rate per year was adjusted so that the total incidence of cancer per lineage over all ages up to 80 years is $m/L = 10^{-9}$, requiring mutation rates for the solid, long-dash, and short-dash curves of, respectively, $v = 10^{-5}$ multiplied by $3.15, 4.35, 8.0$ for all i. (b) An increase in the maximum size of a clone raises peak acceleration until the clone becomes sufficiently large that a mutation is almost certain in a relatively short time period. Parameters as in (a), except that $r_{n-1} = 0.2$, and for the sold, long-dash, and short-dash curves, respectively, $K_{n-1} = 10^6, 10^4, 10^2$, and $v = 10^{-5}$ multiplied by $4.35, 4.45, 6.8$ for all i to keep the total incidence of cancer per lineage at $m/L = 10^{-9}$. (c) Multiple rounds of clonal expansion greatly increase peak acceleration and shift peak acceleration to a later age. Parameters are $n = 4$, $r = 0.5$ for all i, $K_0 = 1$, and $K_{n-1} = 10^6$. For the lower (solid) curve, clonal expansion occurs only in the last round before cancer, so $K_{n-2} = K_{n-3} = 1$. For the middle (long-dash) curve, clonal expansion occurs in the last two rounds before cancer, with $K_{n-2} = 10^6$ and $K_{n-3} = 1$. For the upper (short-dash) curve, clonal expansion occurs in the last three rounds before cancer, with $K_{n-2} = K_{n-3} = 10^6$. The mutation rates for the solid, long-dash, and short-dash curves, respectively, are $v = 5.8 \times 10^{-4}, 9.3 \times 10^{-5}, 1.55 \times 10^{-6}$ for all i to keep the total incidence of cancer per lineage at $m/L = 10^{-5}$. Redrawn from Frank (2004b).

A transition rate might increase rapidly and then not change further. This sudden increase in a transition rate would be similar to a sudden abrogation of a rate-limiting step. Apart from a very brief burst in acceleration, the main effect of a sudden knockout would be a decline in acceleration because fewer limiting steps would remain.

DETAILS

In the first model, transition rates increase with advancing age (Frank 2004a). Let $u_j(t) = uf(t)$, where f is a function that describes changes in transition rates over different ages. We will usually want f to be a nondecreasing function that changes little in early life, rises in midlife, and perhaps levels off late in life. In numerical work, one commonly uses the cumulative distribution function (CDF) of the beta distribution to obtain various curve shapes that have these characteristics. Following this tradition, I use

$$\beta(t) = \int_0^{t/T} \frac{\Gamma(a+b)}{\Gamma(a)\Gamma(b)} x^{a-1} (1-x)^{b-1} \, dx,$$

where T is maximum age so that t/T varies over the interval $[0, 1]$, and the parameters a and b control the shape of the curve. The value of $\beta(t)$ varies from zero at age $t = 0$ to one at age $t = T$.

We need f to vary over $[1, F]$, where the lower bound arises when f has no effect, and F sets the upper bound. So, let $f(t) = 1 + (F - 1)\beta(t)$. Figure 6.8 shows examples of how increasing transition rates affect acceleration.

In the second model, the transition rate between certain stages may rise with clonal expansion. Models of clonal expansion have been studied extensively in the past (Armitage and Doll 1957; Fisher 1958; Moolgavkar and Venzon 1979; Moolgavkar and Knudson 1981; Luebeck and Moolgavkar 2002). I describe the particular assumptions used in Frank (2004b), which allow for multiple rounds of clonal expansion. Multiple clonal expansions would be consistent with multistage tumorigenesis being caused by progressive loss of control of cellular birth and death, ultimately leading to excessive cellular proliferation.

I use the following strategy to study clonal expansion. First, assume that all lines start in stage 0 at birth, $t = 0$, and use the initial condition $x_0(0) = 1$ so that $x_i(t)$ is the probability of a line being in stage i at age t. Second, describe the value of $x_i(t)$ by summing all the influx into and outflux from that stage over the time interval $[0, t]$. Third, cells that enter certain stages undergo clonal expansion. Fourth, clonal expansion increases the number of cells at risk for making the transition to the next stage. To account for this, outflux from a stage increases with the size of clones in that stage.

The probabilities of being in various stages based on the influx and outflux from each stage are

$$x_0(t) = D_0(t, 0)$$

$$x_i(t) = \int_0^t u_{i-1}(s) \, x_{i-1}(s) \, D(t, s) \, ds \qquad i = 1, \ldots, n-1$$

$$x_n(t) = \int_0^t u_{n-1}(s) \, x_{n-1}(s) \, ds,$$

where $u_{i-1}(s) x_{i-1}(s)$ is the influx into stage i at time s, and

$$D_i(t, s) = e^{-\int_s^t u_i(z) dz}$$

is the outflux (decay) as of time t of the influx component that arrived at time s. The integration of x_i values over the time interval $[0, t]$ means that all influxes and outfluxes are summed over the whole time period.

The $u_i(t)$ values vary with time because the fluxes depend on clonal expansion, so we need to express the u's in terms of clonal expansion. I use a logistic model to describe clonal growth. If $y_i(t)$ is the size of the clone in the ith stage at time t, then the clone grows according to $\dot{y}(t) = r_i y_i (1 - y_i / K_i)$, where the dot means the derivative with respect to time, r_i is the maximum rate at which the clone increases, and K_i is the maximum size to which the clone grows. Starting with a single cell, the size of the clone after a time period s of clonal expansion follows the well-known solution for the logistic model (Murray 1989):

$$y_i(s) = \frac{K_i e^{r_i s}}{K_i + e^{r_i s} - 1}.$$

The subscripts describe different stages, so that the different stages may have different rates of increase and maximum sizes.

If we assume that transitions between stages occur by somatic mutation, then for each cell that makes the transition into stage i, the total mutation capacity of that cell lineage is the mutation rate per cell, v, multiplied by the clone size, y, so the outflux of that cell lineage from time s to time t is

$$D_i(t,s) = e^{-\int_s^t v_i y_i(\alpha)d\alpha} = \left(\frac{K_i}{K_i + e^{r_i(t-s)} - 1}\right)^{v_i K_i / r_i}.$$

The total rate of outflux from stage i to stage $i+1$ at time t is

$$u_i(t) = v_i \bar{y}_i(t) = v_i \int_0^t u_{i-1}(s) D_i(t,s) y_i(t-s) \, ds / x_i(t).$$

This model is general enough to fit many different shapes of acceleration curves. However, the goal here is not to fit but to emphasize that a few general processes can explain the differences between tissues in their acceleration patterns.

Figure 6.9a illustrates the effect of changing the rate of clonal expansion, r, in a single round of clonal expansion in stage $n - 1$, similar to the model of Luebeck and Moolgavkar (2002). Slower clonal expansion causes the acceleration in cancer to happen more slowly and to be spread over more years, because slow clonal expansion causes a slow increase in the rate at which a lineage acquires the final transition that leads to cancer. A rapid round of clonal expansion effectively reduces by one the number of steps, n, so that for $n = 4$, one round of rapid clonal expansion yields a nearly constant acceleration of $n - 2 = 2$ over all ages (not shown). By contrast, slow clonal expansion often causes a midlife peak in acceleration, as illustrated in the figure.

Figure 6.9b shows that an increase in maximum clone size raises the peak level of acceleration until the clone becomes large enough that a transition almost certainly occurs in a short time interval, after which further clonal expansion does not increase the rate of progression.

Figure 6.9c shows that multiple rounds of clonal expansion can greatly increase the peak acceleration of cancer. The curves from bottom to top have one, two, or three rounds of clonal expansion.

CONCLUSIONS

Transition rates that increase slowly over time cause acceleration to rise to a midlife peak and then decline late in life. Clonal expansion may be one way in which transition rates rise slowly over time. Alternatively, somatic mutation rates may increase as various checks on the cell cycle and DNA integrity decay with age.

6.6 Summary

This chapter developed the basic models of cancer dynamics under the assumption of multistage progression. Topics included multiple lines of progression and variable rates of transition between stages. The next chapter continues to develop the theory, with emphasis on multiple pathways of progression, genetic and environmental heterogeneity, and a comparison of my models of cancer dynamics with some classical models of aging and of chemical carcinogenesis.

7 Theory II

This chapter continues to develop the quantitative theory of cancer progression and incidence.

The first section analyzes multiple pathways of progression in a particular tissue, in which more than one sequence of events leads to cancer. With multiple pathways, a fast sequence with relatively few steps would dominate incidence early in life and keep acceleration low, whereas a sequence with more steps would dominate incidence later in life and raise the acceleration. Such combinations of sequences can cause the aggregate pattern of incidence to have rising acceleration through midlife, followed by a late-life decline in acceleration.

The second section evaluates how inherited genetic variation affects incidence. Inherited mutations cause individuals to be born with one or more steps in progression already passed. If, in a study, different inherited genotypes cannot be distinguished, then all measurements on cancer incidence combine the incidences of the different genotypes. Rare inherited mutations have little effect on the aggregate incidence pattern. Common inherited mutations cause aggregate incidence to shift between two processes. Mutants dominate early in life: aggregate incidence rises early with a relatively low acceleration, because the mutants have relatively few steps in progression. Normal genotypes dominate later life: aggregate incidence accelerates more sharply with later ages, because the wild type has more steps in progression.

If different genotypes can be distinguished, then one can test directly the role of particular genes by comparison of mutant and normal patterns of incidence and acceleration. The change with age in the ratio of wild-type to mutant age-specific incidence measures the difference in acceleration between the normal and mutant genotype. Under simple models of progression dynamics, the observed difference in acceleration provides an estimate for the difference in the number of rate-limiting stages in progression.

The third section continues study of heterogeneity in predisposition, focusing on continuous variation caused by genetic or environmental factors. Continuous variation may arise from a combination of many genetic variants each of small effect and from diverse environmental factors. I develop the case in which variation occurs in the rate of progression, caused for example by inherited differences in DNA repair efficacy or by different environmental exposures to mutagens.

Populations with high levels of variability have very different patterns of progression when compared to relatively homogeneous groups. In general, increasing heterogeneity causes a strong decline in the acceleration of cancer. To understand the distribution of cancer, it may be more important to measure heterogeneity than to measure the average value of processes that determine rates of progression.

The fourth section relates my models of progression and incidence to the classic Gompertz and Weibull models frequently used to summarize age-specific mortality. The Gompertz and Weibull models simply describe linear increases with age in the logarithm of incidence. Those models make no assumptions about underlying process. Instead, they provide useful tools to reduce data to a small number of estimated parameters, such as the intercept and slope of age-specific incidence.

Data reductions according to the Gompertz and Weibull models can be useful descriptive procedures. However, I prefer to begin with an explicit model of progression dynamics and derive the predicted shape of the incidence curve. Explicit dynamical models allow one to test comparative hypotheses about the processes that influence progression. I show that the simplest explicit models of progression dynamics yield incidence curves that often closely match the Weibull pattern.

The final section reviews applications of the Weibull model to dose-response curves in laboratory studies of chemical carcinogenesis. Most studies fit well to a model in which incidence rises with a low power of the dosage of the carcinogen and a higher power of the duration of carcinogen exposure. Quantitative evaluation of chemical carcinogens provides a way to test hypotheses about the processes that drive progression.

7.1 Multiple Pathways of Progression

Précis

Cancer in a particular tissue may progress by different pathways. Ideally, one would be able to measure progression and incidence separately for each pathway. In practice, observed incidence arises from combined progression over all pathways in a tissue. In this section, I analyze incidence and acceleration when aggregated over multiple underlying pathways of progression.

If one pathway progresses rapidly and another slowly, then incidence and acceleration will shift with age from dominance by the early pathway to dominance by the late pathway. For example, the early pathway may have few steps and low acceleration, whereas the late pathway may have many steps and high acceleration. Early in life, most cases arise from the early, low-acceleration pathway; late in life, most cases arise from the late, high-acceleration pathway.

In this example, the aggregate acceleration curve may be low early in life, rise to a peak in midlife when dominated by the later pathway, and then decline as the acceleration of the later pathway decays with advancing age. Aggregated pathways provide an alternative explanation for midlife peaks in acceleration. In the Conclusions at the end of this section, Figure 7.1 illustrates the main points and provides an intuitive sense of how multiple pathways affect incidence and acceleration. (Various multipathway models are scattered throughout the literature. See the references in Mao et al. (1998)).

DETAILS

For a particular tissue, I assume k distinct pathways to cancer indexed by $j = 1, \ldots, k$. Each pathway has n_j transitions and $i = 0, \ldots, n_j$ states. The probability of being in state i of pathway j at age t is $x_{ji}(t)$. A tissue is subdivided into L distinct lines of progression. A line might be a stem cell lineage, a compartment of the tissue, or some other architecturally defined component. Each line is an independent replicate of the system with all k distinct pathways.

Cancer arises if any of the Lk distinct pathways has reached its final state. All pathways begin in state 0 such that $x_{j0}(0) = 1$ and $x_{ji}(0) = 0$ for all $i > 0$. I interpret $x_{ji}(t)$ as the probability that pathway j is in state i at time t.

The probability that a particular line progresses to malignancy is the probability that at least one pathway in that line has progressed to the final state,

$$z(t) = 1 - \prod_{j=1}^{k} \left[1 - x_{jn_j}(t)\right]. \tag{7.1}$$

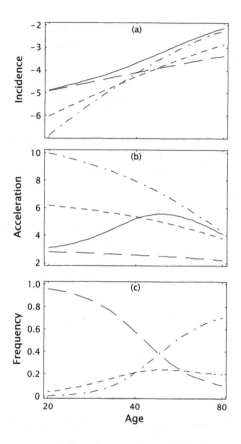

Figure 7.1 Multiple pathways of progression in a tissue influence age-onset patterns of cancer. This figure shows epidemiological patterns for $k = 3$ pathways in a tissue in which there is a single line of progression, $L = 1$. On the y axis, the panels measure (a) log incidence, (b) log-log acceleration (LLA), and (c) frequency of cancer for each pathway. The x axis plots age on a logarithmic scale. The lifetime probability of cancer per individual at age 80 is $m = 0.1$. In each panel, the long-dash curve shows the pathway for which $n_1 = 4$, $u_1 = 0.0103$, and the lifetime probability of cancer is 0.01; the short-dash curve shows the pathway for which $n_2 = 8$, $u_2 = 0.0413$, and the lifetime probability of cancer is 0.02; and the dot-dash curve shows the pathway for which $n_3 = 13$, $u_1 = 0.1016$, and the lifetime probability of cancer is 0.07. The solid curve shows the aggregate over all pathways.

To keep the analysis simple, I focus on k pathways in one line. The solution for multiple lines scales up according to the theory outlined in Section 6.3. Typically, if the total probability of cancer, m, by age T is less than 0.2, then we have $m/L \approx z(T)$, and the cumulative probability of cancer at age t is $p(t) \approx z(t)L$.

The transitions between stages are $u_{ji}(t)$, the rate of flow in the jth pathway from stage i to stage $i + 1$. The transition rates may change with time. These distinct, time-varying rates provide the most general formulation. It is easy enough to keep the analysis at this level of generality, but then we have so many parameters and specific assumptions for each case that it becomes hard to see what novel contributions are made by having multiple pathways. To keep the emphasis on multiple pathways for this section, I assume that all transitions in each pathway are the same, u_j, that transition rates do not vary over time, and that distinct pathways indexed by j may have different transition rates.

Incidence at age t is

$$I = \frac{\dot{z}}{1 - z},$$

where I is the incidence at age t; the numerator, $\dot{z}$, is the total flow into terminal stages at age t; and the denominator, $1 - z$, is proportional to the number of pathways that remain at risk at age t.

The rate of progression for a line is

$$\dot{z} = \sum_{j=1}^{k} \dot{x}_{jn_j} \prod_{i \neq j} (1 - x_{in_i}) = (1 - z) \sum_{j=1}^{k} \frac{\dot{x}_{jn_j}}{1 - x_{jn_j}}.$$

The incidence per pathway is $I_j = \dot{x}_{jn_j}/(1 - x_{jn_j})$, so the previous two equations can be combined to give

$$I = \sum_{j=1}^{k} I_j = \sum_{j=1}^{k} \frac{\dot{x}_{jn_j}}{1 - x_{jn_j}},$$

in words, the total incidence per line is the sum of the incidences for each pathway. Differentiating I yields

$$\dot{I} = \sum_{j=1}^{k} \left(\frac{\ddot{x}_{jn_j}}{1 - x_{jn_j}} + I_j^2 \right).$$

Earlier, I showed that log-log acceleration is $\mathrm{LLA}(t) = t\dot{I}/I$, which can be expanded from the previous expressions.

Using this formula for LLA to make calculations requires applying the pieces from earlier sections. In particular, $\dot{x}_{jn_j} = u_j x_{jn_j-1}$ and $\ddot{x}_{jn_j} =$

$u_j \dot{x}_{jn_j-1} = u_j^2(x_{jn_j-2} - x_{jn_j-1})$. These expansions give everything in terms of x_{ji}, for which we have explicit solutions from an earlier section as

$$x_{ji} = e^{-u_j t} \left(u_j t\right)^i / i! \qquad i = 0, \ldots, n_j - 1 \tag{7.2a}$$

$$x_{jn_j} = 1 - \sum_{i=0}^{n_j-1} x_{ji}. \tag{7.2b}$$

CONCLUSIONS

Figure 7.1 illustrates how multiple pathways affect epidemiological patterns. The pathway marked by the long-dash line in the figure shows a slowly accelerating cause of cancer that dominates early in life. The pathway marked by the dot-dash curve shows a rapidly accelerating cause of cancer that dominates late in life. The aggregate acceleration, shown by the sold curve in Figure 7.1b, is controlled early in life by the slowly accelerating pathway and late in life by the rapidly accelerating pathway. A pathway with intermediate acceleration, shown by the short-dash curve, contributes a significant number of cases through mid- and late life, but does not dominate at any age.

7.2 Discrete Genetic Heterogeneity

Some individuals may inherit mutations that cause them at birth to be one or more steps along the pathway of progression. In this section, I analyze incidence and acceleration when individuals separate into discrete genotypic classes. After deriving the basic mathematical results, I illustrate how genetic heterogeneity affects epidemiological pattern.

PRÉCIS

In the first case, one cannot distinguish between mutant and normal genotypes. If mutated genotypes are rare, then the aggregate pattern of incidence will be close to the pattern for the common genotype. A small increase in cases early in life does develop from the mutated genotypes, but those cases do not contribute enough to change significantly the aggregate pattern.

If the mutants are sufficiently frequent, they may change aggregate acceleration. Early in life, when mutants contribute a significant share of cases, aggregate acceleration may be dominated by the lower acceleration associated with mutants, which have fewer steps in progression than do normal genotypes. Late in life, aggregate acceleration will be dominated by the normal genotype, which has more steps and a higher acceleration. The net effect may be low acceleration early when dominated by the mutants, a rise to a midlife peak as dominance switches to the normal individuals, and a late-life decline in acceleration following the trend set by the normal genotype (Figure 7.2).

In the second case, one can distinguish between mutant and normal genotypes. This is an important case, because it allows one to test directly the role of particular genes by comparison of mutant and normal patterns of incidence and acceleration. I show that the ratio, R, of normal to mutant incidence provides a good way to compare genotypes. The change in this ratio with age on log-log scales is the difference in acceleration between the normal and mutant genotype. Under simple models of progression dynamics, the observed difference in acceleration provides an estimate for the difference in the number of rate-limiting stages in progression.

Details

I assume a single pathway of progression in each line and a single line of progression per tissue, that is, $k = L = 1$. Extensions for multiple pathways and lines can be obtained by following the methods in prior sections. I assume the pathway of progression has n rate-limiting steps, with the transition rate between stages, u. Here, u is the same between all stages and does not vary with time.

A fraction of the population, p_j, has mutations that start them j steps along the pathway of progression; in other words, those individuals have $n - j$ steps remaining before cancer. I refer to individuals that start j steps along as members of class j or as being born in the jth stage of progression.

AGGREGATE PATTERNS

If different genotypes cannot be distinguished, then all measurements on cancer incidence will combine the incidences for the different genotypes. The aggregate rate of transition into the final, cancerous state is $\dot{z} = \sum_{j=0}^{n-1} p_j \dot{x}_{jn-j}$, where x_{ji} is the probability that an individual born in the jth stage has progressed a further i stages. The population-wide

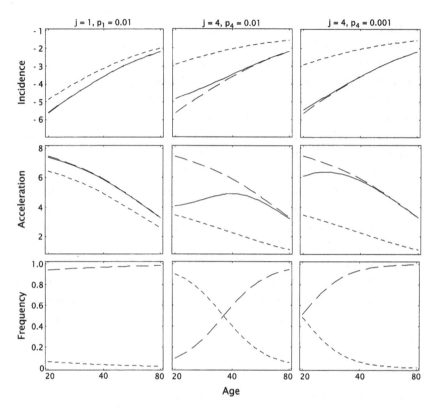

Figure 7.2 Genetic heterogeneity in the population influences aggregate epidemiological patterns. The rows, from top to bottom, are log-log incidence, log-log acceleration, and relative frequency of cancer caused by different genotypes. In each panel, the most common genotype in the population has frequency $p_0 = 1 - p_j$, and a second genotype has frequency p_j, where j is the number of stages in progression by which the mutant genotype is advanced at birth. In all plots, the common genotype has $n = 10$ stages. The long-dash curves show results for the common genotype, the short-dash curves show results for the mutant genotype. The solid curve shows the aggregate pattern for incidence and acceleration. In all plots, the constant rate of transition between stages is $u = 0.0778$ for both the common and mutant genotypes. For all cases, the cumulative probability of cancer at age 80 is approximately 0.1. The rare genotype contributes at most 0.005 to cumulative probability.

cumulative probability of having cancer by age t is $z = \sum_{j=0}^{n-1} p_j x_{jn-j}$. Here, all values of z and x depend on time, but I have dropped the t to keep the notation simple. Eqs. (7.2) provide solutions for x_{ji}, substituting $n - j$ for n_j, and noting the constant transition rates in this section, $u_j = u$ for all j.

From these parts, we can write the total age-specific incidence in the population as

$$I = \frac{\dot{z}}{1-z} = \frac{u \sum_{j=0}^{n-1} p_j \, (ut)^{n-j-1} / \overline{n-j-1!}}{\sum_{i=0}^{n-1} p_i \sum_{j=0}^{n-i-1} (ut)^j / j!},$$

and the log-log acceleration as

$$\text{LLA} = t\dot{I}/I$$

$$= ut \left(\frac{\sum_{j=0}^{n-2} p_j \, (ut)^{n-j-2} / \overline{n-j-2!}}{\sum_{j=0}^{n-1} p_j \, (ut)^{n-j-1} / \overline{n-j-1!}} - \frac{\sum_{i=0}^{n-2} p_i \sum_{j=0}^{n-i-2} (ut)^j / j!}{\sum_{i=0}^{n-1} p_i \sum_{j=0}^{n-i-1} (ut)^j / j!} \right).$$

Figure 7.2 shows that genetic heterogeneity will typically have little effect on aggregate patterns of cancer. That figure assumes a common genotype with $n = 10$ steps and a rare mutant genotype with $n - j$ steps, where j is the number of stages in progression by which the mutant genotype is advanced at birth. If the mutant advances only by $j = 1$, then the patterns differ little between the genotypes. If, however, n is small, as for retinoblastoma, then advancing one step, $j = 1$, can have a significant effect (not shown). Mutants are usually thought to advance progression by just one stage (Knudson 2001; Frank 2005), although relatively little direct evidence exists.

If mutants advance progression by $j = 4$ stages, then the mutants can have a significant impact on aggregate patterns, as shown in the middle column of Figure 7.2 in which the mutant occurs at a frequency of 0.01. However, the mutant must not be too rare—the right column of Figure 7.2 shows that genetic heterogeneity has little effect for $j = 4$, if the mutant occurs at a frequency of 0.001.

COMPARISON BETWEEN GENOTYPES: RATE-LIMITING STEPS

Mutant genotypes may often have little effect on aggregate pattern, as shown in the previous section. However, if one can track the incidence patterns separately for different genotypes, then much can be learned by comparison of incidence patterns between genotypes. Indeed, relative incidence patterns between genotypes may be the most powerful way to learn about cancer progression and the link between particular genes and cancer risk (Knudson 1993, 2001; Frank 2005).

In the next chapter, I will compare retinoblastoma incidence in humans between normal individuals and those who carry a mutation to the retinoblastoma (*Rb*) gene (Section 8.1). I will also compare colon cancer incidence between normal individuals and those who carry a mutation to the *APC* gene. In both cases, the ratio of age-specific incidences between normal and mutant individuals follows roughly along the curve predicted by multistage theory if the mutants begin life one stage further along in progression than do normal individuals (Frank 2005). Here, I develop the theory for predicting the ratio of incidences between normal and mutant genotypes.

Assume a simple model of progression, with n stages and a constant rate of transition between stages, u. Mutant individuals begin life in stage j, and so have $n - j$ stages to progress to cancer. The results of Section 6.2 provide the age-specific incidence for progression through n stages, I_n, so the ratio of incidences of normal and mutant individuals is

$$R = I_n/I_{n-j} = \left(\frac{(ut)^j \, (n - j - 1)!}{(n - 1)!} \right) \left(\frac{S_{n-j-1}}{S_{n-1}} \right), \qquad (7.3)$$

where $S_j = \sum_{i=0}^{j}(ut)^i/i!$. When $j = 1$, then $R \approx ut/(n - 1)$ is often a good approximation (Frank 2005).

When comparing the incidences between two genotypes, it may often be useful to look at the slope of $\log(R)$ versus $\log(t)$, which is

$$\Delta\text{LLA} = \frac{d\log (R)}{d\log (t)} = \frac{d\log (I_n) - d\log \left(I_{n-j}\right)}{d\log (t)}$$

$$= \text{LLA}_n - \text{LLA}_{n-j}$$

$$= j - ut \left(\frac{S_{n-2}}{S_{n-1}} - \frac{S_{n-j-2}}{S_{n-j-1}} \right), \qquad (7.4)$$

where LLA_k, the log-log acceleration for a cancer with k stages, is given in Eq. (6.3). The slope of $\log(R)$ versus $\log(t)$ is equal to the difference in LLA, so I will sometimes refer to this slope as ΔLLA.

When progression causes acceleration to drop at later ages, then the slope of $\log(R)$ tends to decline with age. For example, in Figure 7.3, cancer develops through a single line of progression, $L = 1$. Often, a small number of progression lines tends to cause acceleration to drop at later ages. By contrast, in Figure 7.4, cancer develops through many lines of progression, $L = 10^8$, which keeps acceleration nearly constant across all ages. Consequently, the ratio of incidences has a constant slope equal

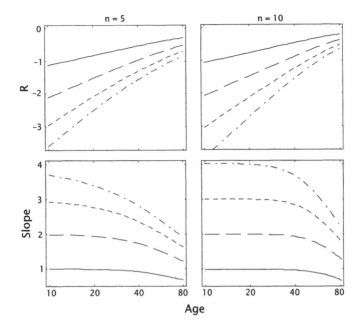

Figure 7.3 Ratio of incidence rates between normal and mutant genotypes when there is a single line of progression, $L = 1$. The normal genotype has n steps in progression to cancer; the mutant has $n - j$ steps. The top row shows the ratio on a $\log_{10}$ scale, calculated from Eq. (7.3). The bottom row shows the slope of the top plots, calculated from Eq. (7.4). The values of j are 1 (solid lines), 2 (long-dash lines), 3 (short-dash lines), and 4 (dot-dash lines). The total incidence for the normal genotype was set to 0.1, which required $u = 0.0304$ for $n = 5$, and $u = 0.0778$ for $n = 10$.

to the number of steps by which a mutation advances progression, that is,

$$\Delta\text{LLA} = \text{LLA}_n - \text{LLA}_{n-j} \approx j. \qquad (7.5)$$

COMPARISON BETWEEN GENOTYPES: TRANSITION RATES

The previous section compared incidence rates between genotypes. In that case, one genotype required n steps to progress to cancer; the other mutant genotype inherited j mutations and began life with only $n - j$ steps remaining. The inherited mutations abrogate rate-limiting steps.

In this section, I make a different comparison. Both genotypes require n steps to complete progression, but the mutant has a higher transition rate between stages. Let the transition rate for the normal genotype be

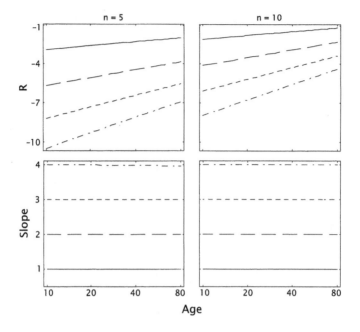

Figure 7.4   Ratio of incidence rates between normal and mutant genotypes when there are multiple lines of progression. For these plots, $L = 10^8$. To keep the cumulative probability at 0.1 for the normal genotype at age 80, $u = 0.00052$ for $n = 5$, and $u = 0.00753$ for $n = 10$. All other aspects match Figure 7.3.

u, and the transition rate for the mutant genotype be $v = \delta u$, with $\delta > 1$. As in Eq. (7.4), I calculate the log-log slope of the ratio of incidences, in this case taking the ratio of mutant to normal genotypes, R. The solution follows from Eq. (6.3):

$$\Delta LLA = LLA_u - LLA_v = ut \left[\frac{\delta S^v_{n-2}}{S^v_{n-1}} - \frac{S^u_{n-2}}{S^u_{n-1}} \right], \qquad (7.6)$$

where $S^\alpha_j = \sum_{i=0}^{j} (\alpha t)^i / i!$

Figure 7.5 illustrates this theory. The left column shows the standard log-log incidence curves. The bottom curve plots the wild-type incidence; the curves above show incidence for mutants with higher transition rates. The right column plots the difference in the slopes of the incidence curves, ΔLLA, between the wild-type and the various mutant genotypes.

The bottom right panel, Figure 7.5h, uses $L = 10^8$ independent lines of progression within the tissue under study. With large L, almost all

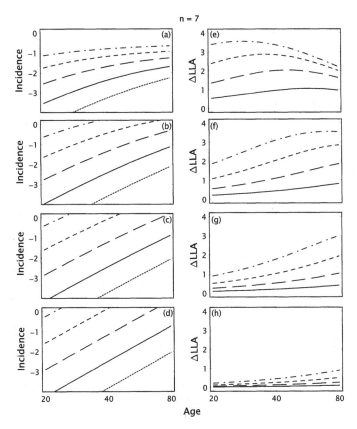

Figure 7.5   Comparison between genotypes with different transition rates. (a-d) The left incidence panels show the standard log-log plot, with incidence on a $\log_{10}$ scale. The bottom, short-dash curve in each incidence panel illustrates the wild-type genotype. The four incidence curves above the wild type show, from bottom to top, increasing transition rates between stages. The transition rate for the bottom curve is u, and for the curves above δu, with $\delta = 6^{i/4}$ for $i = 1, \ldots, 4$. (e-h) The ΔLLA plots on the right show the slope of R, which is the difference between wild-type and mutant genotypes in the slopes of the log-log incidence plots calculated from Eq. (7.6). For example, the solid line in each right panel illustrates the difference in the slopes between the lowest wild-type curve and the solid curve; each line type on the right illustrates the difference in log-log slopes between the wild type and the curve with the matching line type on the left. Each ΔLLA panel has the same parameters as the panel to the left. In each case, the value of u is obtained by solving for the transition rate that yields a cumulative incidence of 0.1 at age 80, where cumulative incidence is given by Eq. (6.5). The values of L from top to bottom are $L = 10^0, 10^2, 10^4, 10^8$.

lineages remain in the initial stage throughout life and have n stages

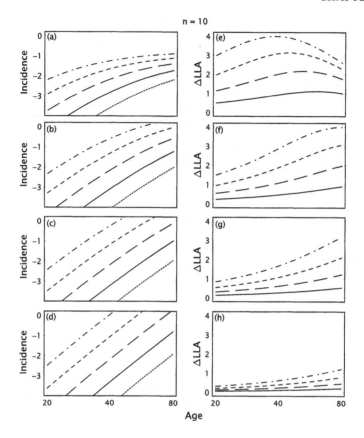

Figure 7.6 Comparison between genotypes with different transition rates. Assumptions are the same as in Figure 7.5, except that $n = 10$ and $\delta = 3^{i/4}$ for $i = 1, \ldots, 4$.

remaining; thus, the log-log incidence slopes remain near $n - 1$ for both wild-type and mutant genotypes.

The top right panel, Figure 7.5e, uses $L = 10^0$ independent lines of progression within the tissue. With small L, the few lineages at risk tend to progress with age through at least the early stages, causing a reduction in the number of remaining stages and a drop in the log-log incidence slope. The mutants, with faster transition rates, advance more quickly through the early stages and so, at a particular age, have fewer stages remaining to cancer. With fewer stages remaining, those mutants have lower log-log incidence slopes, and therefore the difference in slopes, ΔLLA, between wild-type and mutant genotypes increases.

Figure 7.5 uses $n = 7$ stages; Figure 7.6 provides similar plots but with $n = 10$ stages.

In summary, a mutant genotype that increases transition rates will cause a rise in ΔLLA when compared with the wild type. This increase in ΔLLA occurs even though the number of rate-limiting stages is the same for mutant and wild-type genotypes. The amount of the rise with age in ΔLLA depends most strongly on the increase in transition rates caused by the mutant and on the number of independent lines of progression in the tissue.

CONCLUSIONS

The ratio of normal to mutant incidence provides one of the best tests for the role of genetics in progression dynamics. Figures 7.3 and 7.4 show predictions for this ratio under simple assumptions about progression. Similar predictions could be derived by analyzing the ratio of incidences in other models of progression, such as those developed in earlier sections. In Chapter 8, I analyze data on the observed ratio of incidences between normal and mutant genotypes. Those ratio tests provide the most compelling evidence available that particular inherited mutations reduce the number of rate-limiting stages in progression.

7.3 Continuous Genetic and Environmental Heterogeneity

Quantitative traits include attributes such as height and weight that can differ by small amounts between individuals, leading to nearly continuous trait values in large groups (Lynch and Walsh 1998). All quantitative traits vary in populations. With regard to cancer, studies have demonstrated wide variability in DNA repair efficacy (Berwick and Vineis 2000; Mohrenweiser et al. 2003), which influences the rate of progression. Probably all other factors that determine the rate of progression vary significantly between individuals.

Variation in quantitative traits stems from genetic differences and from environmental differences. The genetic side arises mainly from polymorphisms at multiple genetic loci that contribute to inherited polygenic variability. The environmental side includes all nongenetic factors that influence variability, such as diet, lifestyle, exposure to carcinogens, and so on.

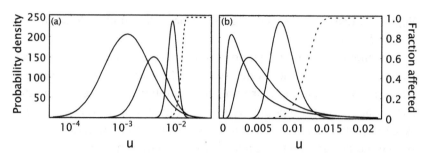

Figure 7.7   The log-normal probability distribution used to describe variation in transition rates, u. (a) In a log-normal distribution of u, the variable $\ln(u)$ has a normal distribution with mean m and standard deviation s. The three solid curves show the distributions used to calculate three of the curves in Figure 7.8. The solid curves from right to left have (m, s) values: $(-4.77, 0.2)$, $(-5.25, 0.6)$, and $(-5.75, 1)$. The dotted line shows the probability that an individual will have progressed to cancer by age 80, measured by the fraction affected on the right scale. I calculated the dotted line using the parameters given in Figure 7.8. (b) Same as panel (a) but with linear scaling for u along the x axis.

In this section, I analyze how continuous variation influences epidemiological pattern. The particular model I study focuses on variation between individuals in the rate of progression. My analysis shows that populations with high levels of variability have very different patterns of progression when compared to relatively homogeneous groups. In general, increasing heterogeneity causes a strong decline in the acceleration of cancer.

PRÉCIS

I use the basic model of multistage progression, in which carcinogenesis proceeds through n stages, and each individual has a constant rate of transition between stages, u. To study heterogeneity, I assume that u varies between individuals. Both genetic and environmental factors contribute to variation.

There are L independent lines of progression within each individual, as described in Section 6.3. I use a large value, $L = 10^7$, which causes loglog acceleration (LLA) to be close to $n - 1$, without a significant decline in acceleration late in life (Figure 6.1).

To analyze variation in transition rates between individuals, I assume that the logarithm of u has a normal distribution with mean m and standard deviation s. This sort of log-normal distribution often occurs

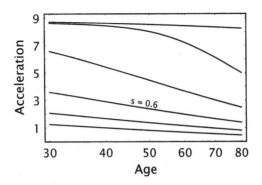

Figure 7.8   Acceleration for different levels of phenotypic heterogeneity in transition rates. Each curve shows the acceleration in the population when aggregated over all individuals, calculated by Eq. (7.9). I used a log-normal distribution for $f(u)$ to describe the heterogeneity in transitions rates, in which $\ln(u)$ has a normal distribution with mean m and standard deviation s. To get each curve, I set a value of s and then solved for the value of m that caused $1-b = 0.1$ of the population to have cancer by age 80 (see Eq. (7.7)). With this calculation, 95% of the population has u values that lie in the interval $(e^{m-1.96s}, e^{m+1.96s})$ (see Figure 7.7). For all curves, I used $n = 10$ and $L = 10^7$. For the curves, from top to bottom, I list the values for (m, s) : low–high, where low and high are the bottom and top of the 95% intervals for u values: $(-4.64, 0) : 0.0097 - 0.0097$; $(-4.77, 0.2) : 0.0057 - 0.013$; $(-5.00, 0.4) : 0.0031 - 0.0015$; $(-5.25, 0.6) : 0.0016 - 0.017$; $(-5.50, 0.8) : 0.00085 - 0.020$; and $(-5.75, 1) : 0.00045 - 0.023$. I tagged the curve with $s = 0.6$ to highlight that case for further analysis in Figure 7.9.

for quantitative traits that depend on multiplicative effects of different genes and environmental factors (Limpert et al. 2001).

Figure 7.7 shows examples of log-normal distributions. Note that a small fraction of individuals has large values relative to the typical member of the population. In terms of cancer, such individuals would be fast progressors and would contribute a large fraction of the total cases.

The question here is: How does heterogeneity influence epidemiological pattern? To study this, I increase variability by raising the parameter s in the log-normal distribution, which increases the variability in transition rates, u. To measure epidemiological pattern, I analyze how changes in s affect log-log acceleration.

Figure 7.8 shows that increasing variability causes a large decline in acceleration when epidemiological pattern is measured over the whole population. In this example, s measures variability: in the top curve, $s = 0$ and the population contains no variability; in the second curve from the top, $s = 0.2$, showing the effect of a small amount of variability; the curves below increase variability with values of $s = 0.4, 0.6, 0.8, 1.0$, respectively.

In Figure 7.8, focus on the curve labeled $s = 0.6$. That curve shows the acceleration of cancer in the total population. Figure 7.9 illustrates the contribution to that aggregate curve by different subgroups of the population with different values of the transition rate, u.

Figure 7.9a plots the contribution of each subgroup in the population: the sum of the individual curves determines the aggregate curve in Figure 7.8. At different ages, each subgroup contributes differently to the aggregate pattern. The solid curve shows the top 2.5% of the population with the highest values of u, defined in the legend as the group between the 97.5th percentile and the 100th percentile. The legend gives the percentile levels for the other curves.

In Figure 7.9a, the solid curve shows that those who progress the fastest contribute most strongly to acceleration early in life. In Figure 7.9b, the solid curve shows the fraction of individuals in that group who have progressed to cancer; already by age 30, ten percent of that group has developed cancer, and by age 60, nearly everyone in that group has progressed.

Returning to Figure 7.8a, we can see that, as age increases, successive groups rise and fall in their contributions to total acceleration in the population. The contribution of each group peaks as the fraction of individuals affected in that group increases above ten percent (Figure 7.9b), and then the contribution declines as nearly all individuals in the group progress to cancer.

Figure 7.9c shows the acceleration pattern if each subgroup were itself the total population. Each group is itself heterogeneous, but with variation over a smaller scale than in the aggregate population. The acceleration pattern is relatively high and constant within all groups except the two highest groups, comprising 5% of the population, who progress very fast.

Figure 7.9b shows that under heterogeneity, cancer forms a rather sharp boundary between those strongly prone to disease, who progress with near certainty, and those less prone, who progress with low probability. This kind of sharp cutoff between those affected and those who escape is sometimes called truncation selection.

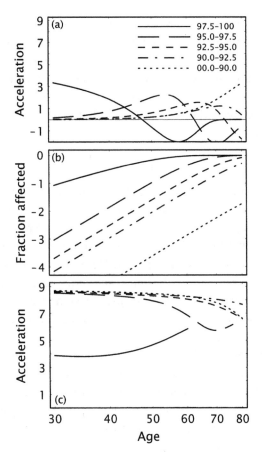

Figure 7.9 Explanation of the drop in aggregate acceleration caused by population heterogeneity. Each panel shows patterns for different segments of the
population stratified by transition value, u. The legend in (a) shows that each of
the first four strata comprise 2.5% of the population, that is, the top 2.5% of u
values, the next 2.5%, and so on. The fifth stratum includes the rest of the population, with individuals that have u values that fall between the 0th and 90th
percentiles. All panels have parameters that match the curve labeled $s = 0.6$
in Figure 7.8. (a) The contribution of each stratum to the aggregate LLA of the
population. I calculated each curve from Eq. (7.8), with denominators integrated
over all values of u, and numerators integrated over values of u within each stratum and divided by the total probability contained in the stratum. Total LLA
equals the sum of the curves. (b) Fraction of individuals within each stratum
who suffer cancer by age 80, calculated as $1 - b$ in Eq. (7.7), integrated over u
values within each stratum, and divided by the total probability contained in
the stratum. (c) LLA calculated within each stratum by integrating numerators
and denominators in Eq. (7.8) over values of u within each stratum.

The truncating nature of selection in this example can also be seen in Figure 7.7, in which the dotted line measures the probability that an individual will have progressed to cancer by age 80 (right scale). Those few individuals with higher u values progress with near certainty; the rest, with lower u values, rarely progress to cancer. The transition is fairly sharp between those values of u that lead to cancer and those values that do not.

DETAILS

I assume a single pathway of progression in each line, $k = 1$, and allow multiple lines of progression per tissue, $L \geq 1$. Extensions for multiple pathways can be obtained by following the methods in earlier sections. I assume the pathway of progression has n rate-limiting steps with transition rate between stages, u. Here, u is the same between all stages and does not vary with time. Each individual in the population has a constant value u in all lines of progression. The value of u varies between individuals. In this case, u is a continuous random variable with probability distribution $f(u)$.

I obtain expressions for incidence and log-log acceleration that account for the continuous variation in u between individuals. To start, let the probability that a particular line of progression is in stage i at time t be $x_i(t, u)$, for $i = 0, \ldots, n$. For a fixed value of u, we have from Section 6.2 that $x_i(t, u) = e^{-ut}(ut)^i/i!$ for $i = 0, \ldots, n-1$ and $x_n(t, u) = 1 - \sum_{i=0}^{n-1} x_i(t, u)$.

The probability that an individual has cancer by age t is the probability that at least one of the L lines has progressed to stage n, which from Eq. (6.5) is

$$p(t, u) = 1 - [1 - x_n(t, u)]^L.$$

Incidence is the rate at which individuals progress to the cancerous state divided by the fraction of the population that has not yet progressed to cancer. The rate at which an individual progresses is $\dot{p}(t, u)$, the derivative of p with respect to t. To get the average rate of progression over individuals with different values of u, we sum up the values of $\dot{p}(t, u)$ weighted by the probability that an individual has a particular value of u. In the continuous case for u, we use integration rather than summation, giving the average rate of progression in the population as

$$a = \int \dot{p}(t, u) f(u) \, du.$$

The fraction of the population that has not yet progressed to cancer is

$$b = 1 - \int p(t, u) f(u) \, du, \tag{7.7}$$

which is one minus the average probability of progression per individual.

With these expressions, incidence is $I(t) = a/b$, and log-log acceleration is

$$\text{LLA}(t) = \frac{d \log(I)}{d \log(t)} = t(\dot{I}/I) = t\left(\frac{\dot{a}}{a} - \frac{\dot{b}}{b}\right). \tag{7.8}$$

Because $\dot{b} = -a$, we can also write

$$\text{LLA}(t) = t\left(\frac{\dot{a}}{a} - \frac{\dot{b}}{b}\right) = t\left(\frac{\dot{a}}{a} + \frac{a}{b}\right) = t\left(\frac{\dot{a}}{a} + I\right). \tag{7.9}$$

To make calculations, we need to express a and $\dot{a}$ in terms of x_i, for which we have explicit solutions. First, to expand a, we need $\dot{p} = L\dot{x}_n(1 - x_n)^{L-1}$, with $\dot{x}_n = ux_{n-1}$ (see Eqs. 6.1). Second, $\dot{a} = \int \ddot{p}(t, u) f(u) du$, with $\ddot{p} = L[\ddot{x}_n(1 - x_n)^{L-1} - \dot{x}_n^2(L - 1)(1 - x_n)^{L-2}]$ and $\ddot{x}_n = u\dot{x}_{n-1} = u^2(x_{n-2} - x_{n-1})$.

Conclusions

Increasing heterogeneity causes a strong decline in the acceleration of cancer. Heterogeneity could, for example, cause a cancer with $n = 10$ stages to have acceleration values below 5 that decline with age. Thus, low values of acceleration (slopes of incidences curves) do not imply a limited number of stages in progression. Heterogeneity must be nearly universal in natural populations, so heterogeneity should be analyzed when trying to understand differences in epidemiological patterns between populations.

Heterogeneity in progression rates causes cancer to be a form of truncation selection, in which those above a threshold almost certainly develop cancer and those below a threshold rarely develop cancer. Under truncation selection, the amount of variation in progression rates will play a more important role than the average rate of progression in determining what fraction of the population develops cancer and at what ages they do so. To understand the distribution of cancer, it may be more important to measure heterogeneity than to measure the average value of processes that determine rates of progression.

7.4 Weibull and Gompertz Models

PRÉCIS

Demographers and engineers use Weibull and Gompertz models to describe age-specific mortality and failure rates. A simple form of the Weibull model assumes that failure rates versus age fit a straight line on log-log scales. This matches the simplest multistage model of progression dynamics under the assumption that log-log acceleration remains constant over all ages.

The advantage of the Weibull model is that it makes no assumptions about underlying process, and allows one to reduce data description to the two parameters of slope and intercept that describe a line. Comparison between data sets can be made by comparing the slope and intercept estimates.

The disadvantage of the Weibull model is that, because it is a descriptive model that makes no assumptions about underlying process, one cannot easily test hypotheses about how particular factors affect the processes of progression. I prefer an explicit underlying model of progression dynamics. In some cases, such as the simplest multistage model, the solution based on explicit assumptions about progression leads to an approximate Weibull model.

The common form of the Gompertz model arises by assuming a constant value for the slope of incidence versus age on log-linear scales: that is, logarithmic in incidence and linear in age. The advantages and disadvantages for the Weibull model also apply to the Gompertz model.

DETAILS

The Weibull model describes age-specific failure rates. Engineers use the Weibull model to analyze time to failure for complex control systems, particularly where system reliability depends on multiple subcomponents. Multicomponent failure models have a close affinity to multistage models of disease progression. Demographers also use the Weibull model to describe the rise in age-specific mortality rates with increasing age.

Both engineers and demographers have observed that the Weibull model provides a good description of age-specific failure rates in many situations, so they use the model to fit data and reduce pattern description to a few simple parameters. Various forms of the Weibull model exist. A simple and widely applied form can be written as

$$W(t) - W(0) = \alpha t^{\beta},$$

where $W(t)$ is the Weibull failure rate at age t, $W(0)$ is the baseline failure rate, and α and β are parameters that describe how failure rate increases with age.

The simple model of multistage progression with equal transition rates, given in Eq. (6.2), can be rewritten as

$$
\begin{aligned}
I(t) &= \alpha t^{\beta}/S_{n-1} \\
&\approx \alpha t^{\beta} \qquad \text{if } S_{n-1} \approx 1
\end{aligned}
$$

where $\alpha = u^n/(n-1)!$, the exponent $\beta = n-1$, and S_{n-1} is the probability that a particular line of progression has not reached the final disease state by age t.

If $I(t) \approx \alpha t^{\beta}$ is a good approximation of the observed pattern of age-specific incidence, then multistage progression dynamics approximately follows the Weibull model. On a log-log scale, the relation is

$$\log(I) \approx \log(\alpha) + \beta \log(t).$$

With this form of the model expressed on a log-log scale, estimates for the height of the line, $\log(\alpha)$, and the slope, β, provide a full description of the relation between incidence and age. The log-log acceleration for this pattern of incidence is β, the slope of the line.

Whenever log-log acceleration remains constant with age, the multistage and Weibull models will be similar. The previous sections discussed the assumptions under which log-log acceleration remains constant with age.

The Weibull model simply describes pattern, and so cannot be used to develop testable predictions about the processes that control age-specific rates. With multistage models of progression, we can predict how incidence will change in individuals with inherited mutations compared with normal individuals, or how incidences of different diseases compare based on the number of stages of progression, the number of independent lines of progression, the variation in transition rates between stages, and the temporal changes in transition rates over a lifetime.

The Gompertz model provides a widely used alternative description of mortality rates. Let $G(t)$ be the age-specific mortality rate of a Gompertz model, and let a dot denote the derivative with respect to t. The Gompertz model assumes that the mortality rate increases at a constant rate y with age:

$$\dot{G} = yG.$$

Solving this simple differential equation yields

$$G(t) = ae^{yt},$$

where $a = G(0)$. From the differential equation, we can also write

$$\frac{\dot{G}}{G} = \frac{d\ln(G)}{dt} = y,$$

which shows that the slope of the logarithm of mortality rate with respect to time is the constant y. Horiuchi and Wilmoth (1997, 1998) defined $d\ln(G)/dt$ as the life table aging rate.

The Gompertz model arises when one assumes a constant life table aging rate. As with the Weibull model, the Gompertz model describes the pattern that follows from a simple assumption about age-related changes in failure rates. Neither model provides insight into the processes that influence age-related changes in disease. However, these models can be useful when analyzing certain kinds of data. For example, the observed age-specific incidence curves may be based on relatively few observations. With relatively few data, it may be best to estimate only the slope and intercept for the incidence curves and not try to estimate nonlinearities.

When fitting a straight line on a log-log scale, one is estimating Weibull parameters. Similarly, fitting a straight line of incidence versus time on a log-linear scale estimates parameters from a Gompertz model. The Weibull distribution may be the better choice because it provides a linear approximation to an underlying model of multistage progression dynamics.

CONCLUSIONS

Weibull and Gompertz models provide useful tools to reduce data to a small number of estimated parameters. However, I prefer to begin with an explicit model of progression dynamics and derive the predicted shape of the incidence curve. Explicit dynamical models allow one to test comparative hypotheses about the processes that influence progression.

7.5 Weibull Analysis of
Carcinogen Dose-Response Curves

PRÉCIS

Peto et al. (1991) provided the most comprehensive experiment and analysis of carcinogen dose-response curves. In their analysis, they compared the observed age-specific incidence of cancer (the response) over varying dosage levels. They described the incidence curves by fitting the data to the Weibull distribution. They also related the Weibull incidence pattern to the classic Druckrey formula for carcinogen dose-response relations. The Druckrey formula summarizes the many carcinogen experiments that give linear dose-response curves when plotting the median time to tumor onset versus dosage of the carcinogen on log-log scales (Druckrey 1967).

I discussed the Druckrey equation, the data from Peto et al.'s study, and some experimental results from other carcinogen experiments in Section 2.5. Here, I summarize the theory that ties the Weibull approximation for incidence curves to the Druckrey equation between carcinogens and tumor incidence.

DETAILS

Define the instantaneous failure rate as $\lambda(t)$. Cumulative failure intensity is $\mu(t) = \int_0^t \lambda(x)dx$. Then, from the nonstationary Poisson process, the probability of survival (nonfailure) to age t is

$$S(t) = e^{-\mu(t)}$$

and failure is $1 - S$.

Note that median time to failure, m, is

$$S(m) = 0.5 = e^{-\mu(m)}$$

and so

$$\ln(0.5) = -\mu(m).$$

Age-specific incidence, $I(t)$, is the instantaneous decrease in survival divided by the fraction of the original population still surviving, thus

$$I(t) = -S'/S = -d\ln(S)/dt = \lambda(t),$$

so the instantaneous failure rate from the nonstationary Poisson process is also the age-specific incidence rate.

Cumulative incidence sums up the age-specific incidences; cumulative incidence measures the total failure intensity over the total time period, thus

$$CI(t) = \int_0^t I(x)\,dx = \int_0^t -d\ln(S(x))/dx$$

$$= -\ln(S(t)) = \int_0^t \lambda(x)\,dx = \mu(t).$$

This background provides the details needed to decipher the rather cryptic analysis in Peto et al. (1991) on the Weibull distribution and the Druckrey equation.

To start, assume that cumulative failure follows the Weibull distribution

$$\mu(t) = -\ln(S) = bt^n.$$

Then the median time to failure is

$$\mu(m) = -\ln(0.5) = bm^n$$

and so

$$b = -\ln(0.5)/m^n$$

and

$$CI = \mu(t) = bt^n = -\ln(0.5)(t/m)^n.$$

Thus, the median, m, and the exponent, n, completely determine the course of survival, time to failure, and incidence.

For carcinogen experiments, Druckrey and others have noted an excellent linear fit on log-log scales between the median time to tumor, m, and dosage, d, such that

$$\log(m) = k_1 - (1/s)\log(d),$$

which means that, in the form usually given in publications,

$$k_1 = dm^s.$$

To use these empirical relations in the incidence formulae above, where patterns depend on t^n and on m, we can use $s = n/r$, thus

$$k_1 = dm^{n/r}$$

and

$$m = (k_1/d)^{r/n}.$$

Substituting for m in our previous formulae,

$$CI = \mu(t) = \frac{-\ln(0.5)\, d^r t^n}{k_1^r} = k_2 d^r t^n,$$

which suggests that cumulative incidence depends on the rth power of dose and the nth power of age, with k values fit to the data.

Note that if $d = 0$, this formula for incidence suggests no cancer in the absence of carcinogen exposure. If there is a moderate to high dosage, then almost all cancers will be excess cases induced by carcinogens. However, one may wish to correct for background cases, either by interpreting CI as excess incidence or by substituting $(d + \delta)^r$ for d, where $\delta > 0$ explains the background cases.

Conclusions

This section provided the technical details to analyze experimental studies of carcinogens. Those studies measure the relation between tumor incidence and age at different dosage levels. The analysis then estimates the effect of dosage on the time to tumor development. Most studies fit well to a model in which the cumulative incidence up to age t rises with $d^r t^n$, where d is dose, t is age, the exponent r is the log-log slope for incidence versus dosage, and n is the log-log slope for cumulative incidence versus age.

7.6 Summary

A wide variety of incidence and acceleration curves can be drawn based on reasonable assumptions about progression and heterogeneity. That great flexibility of the theory means that it is easy to fit a model to observations. A theory that fits almost any observable pattern explains little; insights and testing of ideas cannot come from simply fitting the theory to observations.

The value of the theory arises from comparative hypotheses. The models predict how incidence and acceleration change between groups with different genotypes or different exposures to carcinogens. If one can consistently predict how perturbations to certain processes shift incidence and acceleration, then one has moved closer to understanding the processes of carcinogenesis. The following chapters describe comparative studies.

8   Genetics of Progression

Genes affect cancer to the extent that they alter age-specific incidence. Thus, the most powerful empirical analysis compares age-specific incidence between normal and mutated genotypes. This chapter describes comparative studies between genotypes.

The first section compares mutant and normal genotypes in human populations. I begin with the classic study of retinoblastoma. An inherited mutation in the *Rb* gene causes a high incidence of bilateral retinal tumors. Individuals who do not inherit a mutation suffer rare unilateral tumors. The age-specific acceleration of unilateral cases is one unit higher than the acceleration of bilateral cases, consistent with the prediction that most of the individuals who suffer bilateral retinoblastoma were born advanced by one stage in progression because of an inherited mutation.

A similar comparison between inherited and sporadic cases of colon cancer shows that the sporadic cases have an acceleration approximately one unit greater than inherited cases. The decrease in acceleration for individuals who inherit a mutation to the *APC* gene supports the hypothesis that such mutations cause their carriers to be born one stage advanced in progression.

The second section compares incidence between different genotypes in laboratory animals. The controlled genetic background makes clearer the causal role of particular mutations in shifting age-specific incidence. I describe the quantitative methods needed to test hypotheses with the small sample sizes commonly obtained in lab studies. I then present a full analysis of one example: the change in age-specific incidence and acceleration between four genotypes with different knockouts of DNA mismatch repair genes. Knockouts that cause a greater increase in mutation rate had earlier cancer onset and a lower age-specific acceleration. The lower acceleration suggests some hypotheses about how the mismatch repair mutations affect the rate of cancer progression.

The third section compares breast cancer incidence between human groups classified by the age at which a first-degree relative developed the disease. The earlier the age of onset for the affected first-degree relative, the faster the rate of progression. Those who progressed more quickly appeared to have an inherited polygenic predisposition. Greater polygenic predisposition was associated with lower age-specific acceleration. I discuss various hypotheses about why such predisposition may increase incidence and reduce acceleration.

8.1 Comparison between Genotypes in Human Populations

Comparisons between sporadic and inherited cancers provide powerful support for multistage theory. With new genomic techniques, comparison of age-specific incidence between human groups with different genotypes will become increasingly easy to accomplish. So, it is important to have a clear sense of what has already been done and what can be learned in the future.

RETINOBLASTOMA

Bilateral retinoblastoma, in which tumors develop in both eyes, is an inherited disease. Most unilateral cases occur sporadically. Knudson (1971) predicted that bilateral cases follow age-specific patterns consistent with one inherited mutation (hit) and the need for only one somatic hit to produce a tumor. By contrast, Knudson predicted that unilateral cases require two somatic hits to form a tumor.

Figure 8.1 compares age-specific incidence of bilateral (inherited) and unilateral (sporadic) cases. The typical measure of age-specific incidence is the number of cases in an age group divided by the number of persons at risk in that age group. However, given the small sample sizes and the difficulty of measuring the base population that represents the number of persons at risk, Knudson analyzed incidence as the number of cases not yet diagnosed at a particular age divided by the total number of cases eventually diagnosed, in other words, the fraction of cases not yet diagnosed.

Knudson (1971) fit the bilateral cases to the model $\log(S) = -k_1 t$, where S is the fraction of cases not diagnosed, k_1 is a parameter used to fit the data, and t is age at diagnosis. He fit the unilateral cases to the model $\log(S) = -k_2 t^2$, where k_2 is a parameter used to fit the data. The figure shows a reasonable fit for both models, with $k_1 = 1/30$ and $k_2 = 4 \times 10^{-5}$.

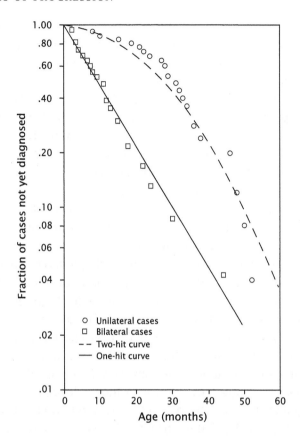

Figure 8.1   Incidence of unilateral and bilateral retinoblastoma. Redrawn from Knudson (1971).

Knudson (1971) gave various theoretical justifications for why inherited and sporadic forms should follow these simple models of incidence, proportional either to t for one hit or t^2 for two hits. However, his theoretical arguments in that paper ignored the way in which the retina actually develops. In a later pair of papers, Knudson and his colleagues produced a theory of incidence that accounts for retinal development (Knudson et al. 1975; Hethcote and Knudson 1978).

Consider, for example, an individual who inherits one mutation. All dividing cells in the retina that are at risk for transformation can be transformed by a single additional somatic mutation. As the retina grows, the number of cells at risk for a somatic mutation increases, causing a rise in risk with age. However, the retina grows to near its final number of cells by around 60 months of age, causing cell division to slow and reducing the risk per cell with age. Change in overall risk with age depends on the opposing effects of the rise in cell number and the decline in the rate of cell division.

Hethcote and Knudson (1978) developed a mathematical theory based on cellular processes of retinal development, and fit their model to an extended set of data on inherited and sporadic retinoblastoma. The basic pattern in the data remains the same as in Figure 8.1, but the later model fits parameters for the somatic mutation rate and for aspects of cell population size and cell division rate.

At first glance, the realistic model based on cell populations and cell division may seem attractive. However, many factors affect the incidence of human cancers, including environment, cell-cell interactions, tissue structure, and somatic mutations during different phases of tumor development. No model can account for all of those factors, and so incidence data can never provide accurate estimates for isolated processes such as somatic mutation rate or cell division rate.

Knudson's main insight was simply that age-specific incidence of inherited and sporadic retinoblastoma should differ in a characteristic way if cancer arises by two hits to the same cell. He obtained the data and showed that very simple differences in incidence do occur. The next step is to understand why the observed differences follow the particular patterns that they do. Detailed mathematical theory based on cell division and mutation rate provides insight about the factors involved, but with regard to data analysis, that theory depends too much on the difficult task of estimating parameters of mutation and cell division from highly variable incidence data.

RATIO OF SPORADIC TO INHERITED INCIDENCE

I advocate theory more closely matched to Knudson's original insight and to what one can realistically infer given the nature of the data (Frank 2005). According to Knudson's theory, bilateral tumors arise from single hits to somatic cells with an inherited mutation. The rate at which a hit occurs in the developing retina at a particular age depends on many factors, including the number of target cells and the rate of cell division. But we cannot get good estimates for those factors, so let us use the observations for bilateral cases at different ages to estimate the rate at which a somatic mutation occurs in the tissue at a particular age, subsuming all the details that together determine that rate.

In particular, we take our estimate for age-specific bilateral incidence as our estimate for the rate at which second hits occur in the tissue at a particular age. Clearly, this simplifies the real process; for example, bilateral cases require at least one hit in each eye. However, the probability of two second hits leading to bilateral cases is fairly high at roughly 0.1–0.3 (Figure 2.6c), thus the probability of one second hit is about $\sqrt{0.1}$–$\sqrt{0.3}$, the same order of magnitude as the probability of two second hits. So let's proceed with the simple approach that $I_B(t)$, the incidence of bilateral cases at age t, provides a rough estimate of the rate of second hits to the tissue at age t.

The incidence of unilateral cases can be written as

$$I_U(t) \approx f(t)\, I_B(t),$$

where $f(t)$ is the fraction of somatic cells at age t that carry one somatic mutation, and $I_B(t)$ is approximately the rate at which the second hit occurs and leads to a detectable tumor. The strongest prediction of multistage theory arises from the comparison of sporadic and inherited cases, so we analyze the ratio of unilateral to bilateral incidence at each age:

$$R = \frac{I_U(t)}{I_B(t)} \approx f(t);$$

in words, the ratio of unilateral to bilateral rates should be roughly $f(t)$, the fraction of cells at time t that carry the first hit in individuals that do not inherit a mutation. For example, if $f(t) = 0.1$, then one-tenth of somatic cells have a first mutation, and the susceptibility for sporadic cases is about one-tenth of the susceptibility for inherited cases.

The expected number of somatic mutational events suffered by a gene in a particular cell is the mutation rate per cell division, v, multiplied by the number of cell divisions going back to the embryo. Let the number of cell divisions at age t be $C(t)$, so that $vC(t)$ is the expected number of mutational events. For most assumptions, $vC(t) \ll 1$, so we can take $vC(t) \approx f(t)$ as the fraction of cells at time t that carry a somatic mutation, and thus

$$R = \frac{I_U(t)}{I_B(t)} \approx vC(t). \qquad (8.1)$$

Few attempts have been made to measure the somatic mutation rate per gene per cell division. Yeast provide a convenient model of single eukaryotic cells. For yeast, the mutation rate has been estimated as 10^{-7}-10^{-5} (Lichten and Haber 1989; Yuan and Keil 1990). In mice, Kohler et al. (1991) estimated the frequency of somatic mutations as 1.7×10^{-5}. There are roughly 10^1-10^2 divisions in a mouse cell lineage, so this study suggests a somatic mutation rate per cell division on the order of 10^{-7}-10^{-6}. I use the approximate value of 10^{-6} per gene per cell generation.

The number of cell divisions, $C(t)$, is roughly in the range 15–40, because there are probably about 15–25 cell divisions before the start of retinal development, and it takes about 15 cellular generations to make the $e^{15} \approx 10^6$-10^7 cells in the fully developed retina. Thus, $I_U(t)/I_B(t) \approx 10^{-4}$-$10^{-5}$, and this ratio may increase by a factor of about two during early childhood as $C(t)$ increases from around 15–25 at the start of retinal development to roughly 30–40 in the final cellular generations in the retina.

These rough calculations lead to two qualitative predictions (Frank 2005). First, the ratio of unilateral to bilateral age-specific incidence should be roughly 10^{-4}-10^{-5}. Second, the ratio of unilateral to bilateral incidence should approximately double with age over the period of retinal growth as the number of cellular generations, $C(t)$, increases with time.

Figure 8.2b shows that the ratio of unilateral to bilateral incidence is in the predicted range of 10^{-4}-10^{-5}, roughly the somatic mutation rate multiplied by the number of cellular generations. This ratio approximately doubles from the earliest age of 0–1 to the latest age of 2–3 at which sufficient numbers of bilateral cases occur to estimate incidence rates. The increase of this ratio supports the prediction that unilateral incidence increases relative to bilateral incidence as the number of cellular generations increases.

DIFFERENCE BETWEEN SPORADIC AND INHERITED ACCELERATION

Individuals who inherit a mutation are born one step further along than are individuals who do not inherit a mutation. Thus, my simple theory predicted that the ratio of sporadic to inherited incidence would

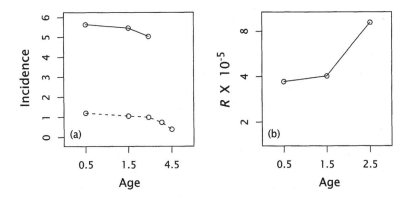

Figure 8.2   Age-specific incidence of retinoblastoma. (a) Bilateral (solid line) and unilateral (dashed line) cases of retinoblastoma per 10^6 population, shown on a $\log_{10}$ scale. Description of the data in Figure 2.6. (b) Ratio, R, of unilateral (I_U) to bilateral (I_B) incidence at each age multiplied by 10^{-5}, using the data in the previous panel. From Frank (2005).

be the probability that nonmutant individuals acquire an extra mutation somatically: approximately the mutation rate per cell division multiplied by the number of cell divisions (Frank 2005). The data shown in Figure 8.2 provide a good match to that prediction when using common assumptions about somatic mutation and cell division.

I now develop a simpler, more general comparative prediction for the difference in incidence between sporadic and inherited cases. In almost all multistage theories, an inherited mutation advances progression and therefore decreases the acceleration of cancer. So, multistage theory predicts that the acceleration of sporadic cases is greater than the acceleration of inherited cases. If we assume that a mutation advances progression by one step, then the theory predicts that the acceleration of inherited cases declines by about one when compared to the acceleration of sporadic cases.

I developed the general theory for comparing accelerations in Sections 7.2 and 7.3. The main features of the theory follow from basic definitions. The ratio of sporadic to inherited incidence is

$$R = \frac{I_S}{I_I}.$$

The slope of R on a log-log scale is

$$\frac{d\log(R)}{d\log(t)} = \frac{d\log(I_S)}{d\log(t)} - \frac{d\log(I_I)}{d\log(t)}.$$

Recall that $d\log(I)/d\log(t)$ is the slope of the incidence curve, or acceleration, when measured on a log-log scale. I called this measure the log-log acceleration (LLA). Thus, the log-log slope of R is the difference in acceleration between sporadic and inherited cases

$$\Delta\text{LLA} = \frac{d\log(R)}{d\log(t)} = \text{LLA}_S - \text{LLA}_I, \qquad (8.2)$$

in which ΔLLA denotes the difference in log-log acceleration.

Figure 8.3 shows the log-log slope of R (ΔLLA) for retinoblastoma, using unilateral incidence to measure sporadic cases and bilateral incidence to measure inherited cases. To calculate the log-log slope of R, I started in Figure 8.3a with the same incidence data as in Figure 8.2a. Estimates for incidence at each age derive from many observations, as described in Figure 2.6. I fit straight lines to the data for unilateral and bilateral cases in Figure 8.3a. The plot in Figure 8.3c shows $\log(R)$, a linearized version of Figure 8.2b. The slope of $\log(R)$ versus $\log(t)$ is about one-half, as shown in Figure 8.3e.

In plotting the retinoblastoma data, the proper scaling for age needs to be considered. So far, I have used age since birth. However, the progression by somatic mutation may begin just after conception. So, it might be reasonable to measure age in years since conception, obtained by adding 0.75 years to age since birth.

The plots in the right column of Figure 8.3 measure age since conception. Using age since conception, the log-log slope in Figure 8.3f is near one, matching the predicted value from simple multistage models, such as in Eq. (6.3). These plots illustrate how incidence data may be used to study the dynamics of cancer progression.

Colon Cancer

Individuals who inherit one mutated copy of the *APC* gene almost invariably develop multiple colon tumors by midlife, causing a disease known as familial adenomatous polyposis (FAP) (Kinzler and Vogelstein 2002). In terms of multistage theory, it may be that individuals with an inherited *APC* mutation begin life one stage further along than do normal individuals (Frank 2005).

Figure 8.4a shows the age-specific incidence for individuals with inherited FAP or noninherited (sporadic) colon cancer. Figure 8.4b plots

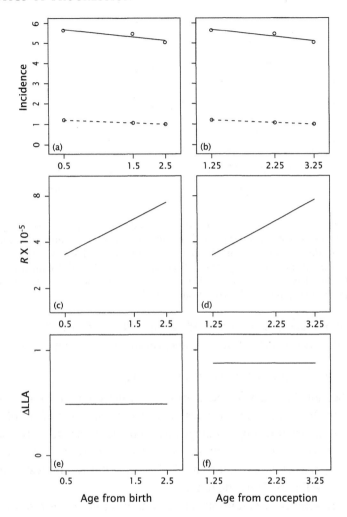

Figure 8.3 Retinoblastoma incidence evaluated with regard to multistage theory. (a and b) Bilateral (solid line) and unilateral (dashed line) incidence of retinoblastoma per 10^6 population, shown on a $\log_{10}$ scale. Description of the data in Figure 2.6. (c and d) Ratio, R, of unilateral (I_U) to bilateral (I_B) incidence at each age multiplied by 10^{-5}, using the fitted lines in the panels above. (e and f) Difference in log-log acceleration between unilateral and bilateral cases, which is the log-log slope of R versus age in Eq. (8.2). The left column shows age from birth; the right column shows age from conception. Ages measured in years. I did not use the unilateral data after age 2.5 years (see Figure 8.2), because retinal cell division slows with age, changing a key process that governs incidence. Without matching data from bilateral cases after 2.5 years, there is no way to calibrate the effect of slowing cell division on the ratio of unilateral to bilateral incidence.

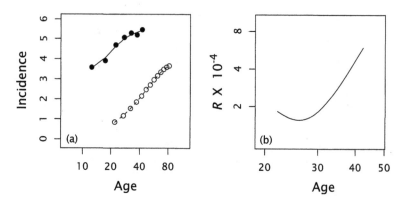

Figure 8.4 Age-specific incidence of inherited and sporadic colon cancer. (a) Inherited colon cancer (FAP) caused by mutation of the *APC* gene (solid circles) and sporadic cases (open circles) per 10^6 population, shown on a $\log_{10}$ scale. Description of the data in Figure 2.6. (b) Ratio of sporadic colon cancer incidence (I_C) to inherited FAP incidence (I_F) at each age multiplied by 10^{-4}, using the data in the previous panel. From Frank (2005).

the ratio of sporadic to inherited age-specific incidence, $R = I_S/I_I$. This ratio increases about 3-fold with age, varying between about 2-6 $\times 10^{-4}$.

RATIO OF SPORADIC TO INHERITED INCIDENCE

In Section 7.2, I developed theory to predict the ratio of age-specific incidence between two genotypes under the assumption of simple step-wise progression through n stages with constant transition rates. One could certainly use more complex models, but there are not enough data to justify particular assumptions. So I stick with the simplest model to see how well it explains the data.

I start with the assumption that sporadic colon cancer requires progression through n stages. Inherited FAP requires progression through only $n - 1$ stages, because at birth those individuals have already advanced by one stage. From Eq. (7.3), we have the ratio of sporadic to inherited cases

$$R \approx \frac{ut}{n - 1},\tag{8.3}$$

noting that the colon has multiple lines of progression, thus the ratio of S_{n-2}/S_{n-1} in Eq. (7.3) will be close to one.

If transitions occur as somatic mutations, then the transition rate per year is the mutation rate per cell division, v, multiplied by the number of cell divisions per year, D, providing the substitution $u = vD$ in Eq. (8.3).

I use $v \approx 10^{-6}$, as discussed in the previous section. The colon epithelium turns over every few days, and stem cells that ultimately renew the tissue probably divide at least once per week, or about $D \approx 50$ times per year. For the number of stages, n, epidemiological and molecular estimates usually fall in the range 4–7 (Armitage and Doll 1954; Fearon and Vogelstein 1990; Luebeck and Moolgavkar 2002). All of these numbers are provisional, but they allow us to predict that the ratio of sporadic to inherited incidence rates should be roughly

$$R \approx \frac{ut}{n-1} = \frac{vDt}{n-1} \approx 10^{-5}t.$$

The data for inherited FAP and sporadic cases can be compared on the range $t = 20$–40, so R is predicted to increase over the range 2–4 × 10^{-4}. Figure 8.4b shows that the ratio of incidences is of the predicted magnitude and increases with age, although the increase with age is slightly greater than predicted.

ACCELERATION IN SPORADIC VERSUS INHERITED CASES

Multistage theory predicts that sporadic cases must progress through at least one more stage than inherited cases. More stages in progression leads to a higher acceleration, so the theory predicts that cases of sporadic colon cancer should accelerate with age more rapidly than the acceleration of inherited cases.

Figure 8.5 shows the same data as in Figure 8.4, with the incidence curves forced to be straight lines. This forced linearity allows an approximate estimate of the log-log slope of R versus t, as shown in Figure 8.5c. The estimated value of 1.5 for this slope is reasonably close to the predicted value of 1, the difference in the number of stages between sporadic and inherited forms.

The theory can be refined in many ways, for example, taking account of the number of independent cell lineages at risk for stepping through the various transition stages. But most reasonable assumptions apply to both the inherited and sporadic rates of transition, and so the ratio of incidence rates remains roughly the same under such refinements.

At present, we have little quantitative information about the different processes that drive progression. Without such details, we get the most insight from simple theories that lead to easily tested comparative predictions. For sporadic versus inherited cancers, two predictions of

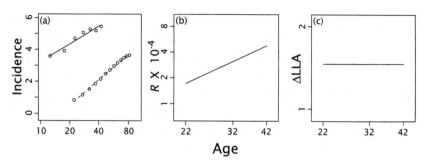

Figure 8.5 Age-specific incidence of inherited and sporadic colon cancer. (a and b) These panels match the corresponding panels in Figure 8.4, with the fitted incidence curves here forced to be linear by assumption. (c) The difference in the log-log acceleration between sporadic and inherited cases, which is the log-log slope of R (see Eq. (8.2)).

multistage theory apply broadly. The first prediction is qualitative: the acceleration of sporadic cases should be greater than the acceleration of inherited cases. The second prediction is quantitative: if inherited cases arise from a single mutation, then the difference in acceleration between sporadic and inherited cases should be about one. My analyses of retinoblastoma and the FAP form of inherited colon cancer support both the qualitative and quantitative predictions.

8.2 Comparison between Genotypes in Laboratory Populations

The previous sections compared the age of cancer onset between individuals with and without particular inherited mutations. Those individuals with inherited mutations progressed more quickly, at a rate consistent with having passed at birth one stage in cancer progression.

Many lab studies with mice or rats compare the age-onset patterns of cancer between different genotypes. Those studies usually focus on whether particular mutations cause faster progression to cancer. In the lab, one can control the environment and use animals that differ only at particular loci. Such studies can provide a strong case for the causal role of certain mutations in cancer progression.

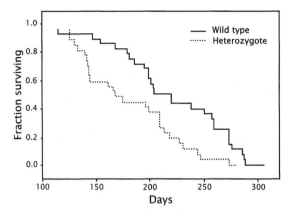

Figure 8.6   Survival of wild-type TRAMP mice versus *Pten* heterozygous TRAMP mice that have one *Pten* allele knocked out. Kwabi-Addo et al. (2001) ascribed death in all 63 mice shown in these plots to either a large primary tumor or to metastatic disease. Survival plots of this sort are often called Kaplan-Meier plots.

Lab studies rarely analyze the quantitative patterns of cancer onset in the way that I did in the previous sections. Instead, the analysis typically emphasizes the qualitative pattern of whether certain combinations of mutations cause earlier or later cancer onset than do other combinations. For example, Figure 8.6 compares the survival of two mouse strains (Kwabi-Addo et al. 2001). One strain has the TRAMP genotype that predisposes mice to develop prostate cancer. The other strain carries the same genes that predispose to prostate cancer, but also is heterozygous at the *Pten* locus, with one allele knocked out. *Pten* mutations are common in many cancers, including cancers of the prostate. The figure shows that the *Pten* heterozygotes progress more rapidly to cancer.

Experimenters usually plot results from these studies as the fraction of mice surviving to a particular age, as in Figure 8.6. In this section, I show how to transform such data into age-specific rates of cancer incidence, allowing comparison of relative rates for different treatments. This transformation to age-specific rates allows one to test particular hypotheses about the dynamics of cancer onset with the limited sample sizes typical of lab studies. I illustrate the method by analyzing the age of cancer onset in different DNA mismatch repair genotypes (Frank et al. 2005).

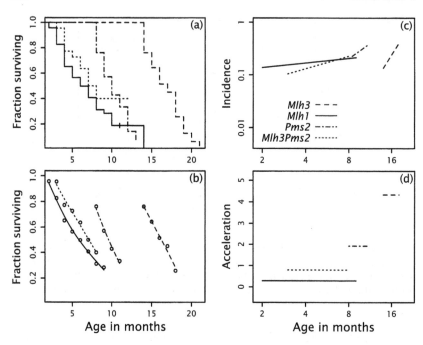

Figure 8.7 Age of lymphoma onset in mice with different mismatch repair genotypes. For each genotype, both alleles at each locus were knocked out. (a) Kaplan-Meier estimate at each age of the fraction of mice that have not yet developed lymphoma among the population of mice that remain at risk. (b) Smoothed curve fit to the estimated survival curve by the smooth.spline function of the R computing language (R Development Core Team 2004) with the smoothing parameter set to 0.5. (c) Incidence of lymphomas on log-log scales. (d) The acceleration of lymphoma onset calculated from the slope of the lines in (c). Redrawn from Frank et al. (2005).

METHODS

Usually, the lab animals in each group have a common genotype and common method of treatment. Each group forms a population in which one observes the time to onset for a particular stage of cancer progression in a particular tissue. From the onset times, one estimates a "survival" curve, where "survival" here means time to onset of some particular event.

In each time interval, for example in each month, one has a listing of how many animals were removed because they suffered the event of interest and how many animals were removed for other causes. If we assume the other causes of removal happen independently of the event of interest, we can use the data to estimate a survival curve.

The Kaplan-Meier survival estimate provides the simplest and most widely used method for lab studies. At each time, t_i, at which events are recorded, the fraction surviving during the interval since the last recording is $\sigma_i = 1 - d_i/n_i$, where d_i is the number of individuals suffering an event since the last time of recording at t_{i-1}, and n_i is the number of individuals at risk during this period. Note that as other causes remove individuals, n_i decreases over time by more than the number of observed events. The fraction of individuals that have not suffered an event (survived) to time t_i is the product of the survival fractions over all time intervals, $S(t_i) = \prod \sigma_j$, where the product of the σ_j's is calculated over all time intervals up to and including t_i.

Figure 8.7 shows the steps by which I transform Kaplan-Meier survival plots (Figure 8.7a) into incidence (Figure 8.7c) and acceleration (Figure 8.7d) plots. These analytical transformations provide an informative way of presenting data with regard to quantitative study of progression dynamics. Frank et al. (2005) give the details for this analysis. Here, I briefly summarize the main points.

The data in Figure 8.7 come from mouse studies of mutant mismatch repair (MMR) genotypes. Defects in the MMR system reduce repair of insertion and deletion frameshift mutations and single base-pair DNA mismatches (Buermeyer et al. 1999). MMR defects can also reduce initiation of apoptosis in response to DNA damage (Edelmann and Edelmann 2004).

I transformed standard survival plots into incidence by first fitting a smoothed curve to the survival data (Figure 8.7b). From the survival curve, $S(t)$, the incidence, measured as probability of death from cancer per month at age t, is

$$I(t) = -\frac{dS(t)}{dt}\frac{1}{S(t)} = -\frac{d\ln(S(t))}{dt}. \tag{8.4}$$

I calculated the incidence curves with Eq. (8.4), put incidence and age on log-log scales, and then fit a straight line through the estimated curves to get the lines of log-log incidence in Figure 8.7c. I fit straight lines because the data provide enough information to get a reasonable estimate of the slope, but not enough information to provide a good estimate of the curvature of the log-log plots at different ages.

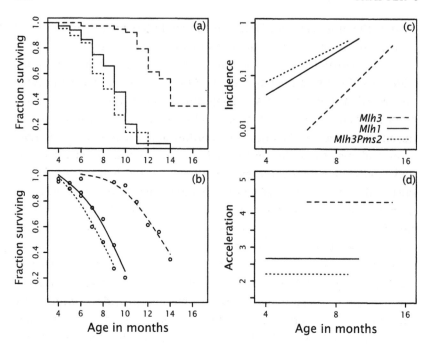

Figure 8.8   Age of gastrointestinal tumor (adenoma) onset in mice with different mismatch repair genotypes. Panels as in Figure 8.7. Redrawn from Frank et al. (2005).

Acceleration in Figure 8.7d shows the slope of the incidence curves. The accelerations are constant over time because I forced the incidence curves to be linear. With more data, one could estimate nonlinear incidence curves, which would allow changes in acceleration with age.

HYPOTHESES AND TESTS

Multistage theory makes three qualitative predictions about the dynamics of cancer. First, the fewer the number of steps in progression that must be passed, the lower the acceleration of cancer with age. In lab experiments, the theory predicts that abrogation of tumor suppressor functions or introduction of oncogenes reduces the acceleration. Second, small to moderate increases in the mutation rate cause greater cancer incidence at earlier ages but do not affect the acceleration. Third, large increases in mutation rate can cause such rapid transitions between stages that certain mutations required for carcinogenesis may no longer limit the rate of tumor formation. If some transitions no longer

Comparison	Type	Acceleration	Age	Mutation
Mlh3 v Mlh3Pms2	GI cancer	+	+	−
Mlh3 v Mlh1	GI cancer	+	+	−
Mlh3 v Mlh3Pms2	Lymphoma	+	+	−
Mlh3 v Mlh1	Lymphoma	+	+	−
Pms2 v Mlh3Pms2	Lymphoma	+	+	−
Pms2 v Mlh1	Lymphoma	+	+	−
Mlh3 v Pms2	Lymphoma	+	+	−

Figure 8.9   Comparison of cancer dynamics for four different mismatch repair genotypes. The '+' and '−' symbols show the direction of change for each comparison. In each comparison, the genotype with the lower mutation rate had a higher acceleration and median age of onset—or, equivalently, the genotype with the higher mutation rate had a lower acceleration and median age of onset. From Frank et al. (2005).

limit the kinetics of carcinogenesis, the number of rate-limiting steps decreases, and the acceleration declines.

MMR genotypes affect both mutation rate and apoptosis in response to DNA damage. Apoptosis suppresses cancer progression and may often be a rate-limiting step in carcinogenesis. Previous work (Chen et al. 2005; Lipkin et al. 2000) showed that the mutation rates for the four knockout genotypes can be ordered as $Mlh3 < Pms2 < Mlh1 \approx Mlh3Pms2$, and decreased apoptosis in response to DNA damage of the four genotypes can be ordered as $Mlh3 \approx Pms2 < Mlh1 \approx Mlh3Pms2$.

Figure 8.9 shows that differences in mutation rate predict the direction of change in acceleration and median age of onset in lymphomas (Figure 8.7) and in gastrointestinal tumors (Figure 8.8). Note that it is possible to have later age of onset and lower acceleration, so acceleration and age of onset are two independent dimensions of the dynamics. The direction of change in mutation rate predicts the direction of change in the acceleration in all 7 cases ($p \approx 0.008$), with the same result for the association between mutation rate and age of onset. Differences in anti-apoptotic effects (not shown) also predict the direction of change in acceleration and age of onset.

Limited sample sizes present the greatest problem in studies that estimate age-specific incidence for particular genotypes. To get around this limitation, I formulated the hypotheses as predictions about the direction of change in comparisons between genotypes. For example, I predicted that acceleration would decline in a sample with relatively stronger defects in mismatch repair when compared against a sample with relatively weaker defects in mismatch repair.

If each key prediction is formulated in a comparative way, laboratory studies with small sample sizes can be used. Each comparison provides a single binary outcome that represents either a success or failure of the theory to predict the direction of change in some attribute of cancer dynamics. The binary outcomes can be aggregated to form a nonparametric test based on the binomial distribution. This allows my approach to be applied to small samples of mice in each genotype. The effective sample size comes from the number of comparisons.

Over the past few years, vast resources have been expended on animal experiments that compare survival curves for different genotypes. If these sorts of experiments were designed and analyzed with dynamics in mind, the research could move to the next level in which the mechanistic consequences of particular genetic pathways are related to the dynamics of carcinogenesis. The data I presented here were not collected to test mechanistic and quantitative hypotheses about dynamics. A simple reanalysis provided significant insights about how DNA repair genotypes affect separately the age of onset and the acceleration of cancer.

8.3 Polygenic Heterogeneity

The previous sections showed how mutations to the mismatch repair genes or *APC* accelerate gastrointestinal cancer, and mutation to *Rb* accelerates retinoblastoma. Those mutations to single genes have simple inheritance patterns and cause major changes in incidence, making them relatively easy to study.

Genetic variation across multiple loci may also strongly affect incidence. However, such polygenic causation creates difficulties in studies, because each particular genetic variant has only a minor effect, shifting incidence by only a small amount.

Comparison of the rate of progression between different genotypes could provide information about the ways in which genetic variants combine to influence cancer incidence. However, if individual genetic variants cause only small changes, then how can one identify genotypes that are sufficiently different with regard to cancer predisposition? One approach is to identify a group of genetically predisposed individuals by studying the first-degree relatives of those who develop cancer early in life. This high-risk group can be compared with a control group of low-risk individuals who do not have an affected first-degree relative.

In a comparison between high- and low-risk groups, two outcomes would suggest polygenic predisposition to cancer. First, the high-risk group must have early onset of cancer as measured by age-specific incidence. Second, one must rule out the possibility that major mutations to single genes, such as *APC*, explain most of the difference in age-specific incidence.

A study of breast cancer showed that those with affected first-degree relatives progress more rapidly than do the controls (Figure 8.10). Interestingly, the earlier the age at which a first-degree relative develops breast cancer, the greater the incidence of those at risk (Peto and Mack 2000).

The slopes of the incidence curves form a set of parallel acceleration curves (Figure 8.11). Those groups whose first-degree relatives had cancer at a relatively earlier age had both greater incidence and lower acceleration. In terms of multistage theory, this negative association between incidence and acceleration arises when the genetically predisposed fast progressors must pass through fewer rate-limiting stages than the slow progressors.

A difference in the number of stages in progression can arise in at least four ways. First, the fast progressors may have genotypes that advance them one or more stages in progression. An advance in initial stage seems to explain the difference in incidence in the single-gene defects, such as retinoblastoma and FAP.

Second, the fast progressors may have less efficient DNA repair and a higher somatic mutation rate, causing progression to advance so rapidly through some stages that those stages are no longer rate-limiting.

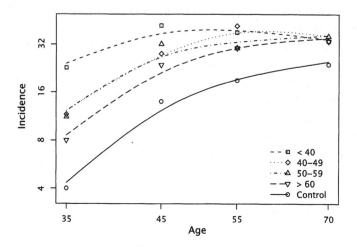

Figure 8.10   Age-specific incidence of breast cancer for individuals with an affected first-degree relative. Incidence shown as cases per 10,000 individuals
per year. The various lines plot the ages at which first-degree relatives were
diagnosed with breast cancer. I calculated incidence from a summary report
on familial breast cancer (Collaborative Group on Hormonal Factors in Breast
Cancer 2001). The report presented data on relative risk for individuals with
affected first-degree kin and on incidence in controls who did not have affected
kin. I calculated incidence as relative risk multiplied by incidence in controls.
The data do not exclude cases in which an affected family carries a major mutation to a gene such as *BRCA1* or *BRCA2*. However, Peto and Mack (2000) used
independent data on the frequency of *BRCA1* or *BRCA2* mutations in affected
individuals of different ages to argue that families carrying major mutations
make up only a small fraction of the total population of families in this study.

Third, the fast progressors may start in the same stage as the slow
progressors and have as many rate-limiting steps to pass but advance
more quickly through stages. At later ages, the fast progressors will
on average have fewer stages remaining. It is the number of stages
remaining that determines acceleration at a particular age (Figure 6.2;
Frank 2004b, 2004d).

Fourth, genetic variants may affect aspects of clonal expansion or
other processes that influence acceleration (Chapters 6 and 7).

Peto and Mack (2000) suggested that incidence reaches a high, constant level after the age at which a first-degree relative develops breast
cancer. Figure 8.10 does show that the incidence curves level off sooner

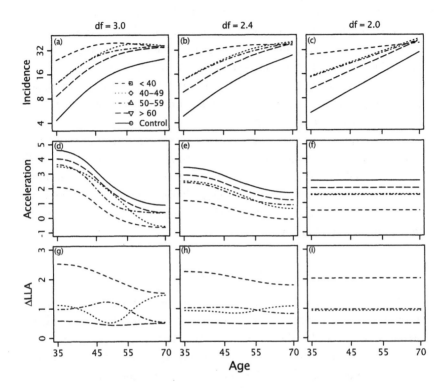

Figure 8.11   Incidence and acceleration of breast cancer in affected families. (a)
This plot is identical to Figure 8.10, with the individual points not shown. Each
curve is derived by fitting a smoothed spline to points at the four ages marked
by ticks on the x axis. In this panel, I used the smooth.spline function of R
with degrees of freedom (df) equal to 3 (R Development Core Team 2004). (b,c)
Incidence curves fit with degrees of freedom equal to 2.4 or 2.0, respectively,
forcing a more linear fit. (d–f) Acceleration, the slope of the incidence curves in
the panels above. The flattening of the acceleration curves near the endpoints
arises at least partly from the spline-fitting procedure, which linearizes the fit
of the incidence curves at the extreme values. (g–i) The differences in the acceleration curves from the panels above; each curve is the difference between the
control curve and the curve for one of the groups with an affected relative. Note
that the accelerations are somewhat erratic because they are derived from the
slope of fitted curves based on observations at only four distinct age categories
(see Figure 8.10). By contrast, the ΔLLA values remain relatively stable under
different smoothing stringencies.

when the first-degree relative is affected at an earlier age. Why might
incidence plateau earlier in faster progressors?

If fast progressors have passed through all but the final stage in cancer progression, and have only one stage remaining, then their annual risk is constant—the risk is just the constant probability of passing the final stage (Frank 2004d). By contrast, families with low genetic risk move through the early stages slowly. In midlife, slow progressors typically have more than one stage to pass, and so continue to have an increasing rate of risk with advancing age.

The key questions concern what sort of genetic variants cause relatively fast or slow progression, and how those genetic variants actually affect the mechanisms and rates of progression. In a later chapter, I discuss genetic variation in more detail. Based on the limited data currently available, one conclusion is that variants in DNA repair efficacy may play an important role.

8.4 Summary

This chapter discussed inherited genetic predisposition to cancer in light of multistage theory. Comparisons between genotypes provide the strongest evidence for the role of particular genes in cancer progression. Indeed, shifts in age-specific incidence may be the only way to measure the consequences of particular genotypes on cancer, and quantitative changes in progression dynamics may be the only way to evaluate the relative importance of particular carcinogenic processes. The next chapter applies the same methods of analysis to chemical carcinogens. The observed shifts of age-specific incidence in response to carcinogens provide a window onto the processes of cancer progression.

9 Carcinogens

Carcinogens shift age-incidence curves. Such shifts provide clues about the nature of cancer progression. For example, a carcinogen that influenced only a late stage in progression would have little effect if applied early in life, whereas a carcinogen that influenced only an early stage would have little effect if applied late in life. By various combinations of treatments, one can test hypotheses about the causes of different stages in progression.

The first section begins with the observation that incidence rises more rapidly with the duration of exposure to a carcinogen than with the dosage. Cigarette smoking provides the classic example, in which incidence rises with about the fifth power of duration and the second power of dosage.

The standard explanation for the relatively weaker effect of dosage compared with duration assumes that a carcinogen affects only a subset of stages. I contrast that standard theory with a variety of alternative explanations. For example, a model in which a carcinogen affects equally all stages also fits the data well. Overall, fitting different models to the data provides little insight.

The second section begins with the observation that lung cancer incidence changes little after the cessation of smoking but increases in continuing smokers. The standard explanation assumes that smoking does not affect the final transition in the sequence of stages of cancer progression. Among those who quit, nearly all subsequent cases arise from individuals who progressed to the penultimate stage while smoking, and await only the final transition. With one stage to go, incidence remains nearly constant over time.

I show once again that a model in which a carcinogen affects equally all stages also fits the data well. Although the data do not distinguish between theories, the various theories do set a basis for connecting how carcinogens influence mechanisms of cellular and tissue change, how those changes affect rates of transition in the stages of tumorigenesis, and how those rates of progression affect incidence curves.

The third section links different mechanistic hypotheses about carcinogen action to predicted shifts in age-incidence patterns. Those links between mechanism and incidence provide a way to test hypotheses about carcinogenic effects on the rate of transition between stages, on the number of stages affected, and on the particular order of affected transitions.

By altering both carcinogen treatment and animal genotype, one may test explicit hypotheses about carcinogenic action. For example, if a carcinogen is believed to cause a particular genetic change, then a knockout of that genotype should be less affected by the carcinogen when measured by age-incidence curves. Such tests can manipulate different components of progression and compare the outcomes to quantitative theories of incidence.

9.1 Carcinogen Dose-Response

Lung cancer incidence increases with roughly the fourth or fifth power of the number of years (duration) of cigarette smoking but with only the first or second power of the number of cigarettes smoked per day (dosage). The stronger response to duration than dosage occurs in nearly all studies of carcinogens. Peto (1977) concluded: "The fact that the exponent of dose rate is so much lower than the exponent of time is one of the most important observations about the induction of carcinomas, and everyone should be familiar with it—and slightly puzzled by it!"

In this section, I first summarize the background concepts and two studies of duration and dosage. I then consider five different explanations. The most widely accepted explanation is that cancer progresses through several stages, causing incidence to rise with a high power of duration, but that a carcinogen usually affects only one or two of those stages in progression, causing incidence to rise with only the first or second power of dosage. However, several alternative explanations also fit the data, so fitting provides little insight. In a later section, I discuss ways to formulate comparative tests. Such comparative tests may help to distinguish between alternative hypotheses and to reveal the processes by which carcinogens influence progression.

Background

In the standard theory, the usual approximation of incidence is $I(t) \approx ku^n t^{n-1}$, where k is a constant, n is the number of rate-limiting transitions between stages that must be passed before cancer, u is the rate of transition between stages, and t is age. Suppose a carcinogen increases the rate of transition between some of the stages to $u(1 + bd)$, where d is dosage and b scales dose level into an increment in transition rate.

If the carcinogen affects r of the transitions, then $I(t) \approx ku^n(1 + bd)^r t^{n-1}$. Two further changes to this equation provide a more useful formula for studies of dosage and duration.

First, in examples such as cigarette smoking, the onset of carcinogen exposure does not begin at birth but at some age t_0 at which smoking starts, so the duration of exposure is $t - t_0 = \tau$.

Second, in empirical studies, one cannot directly estimate u, the baseline transition rate between stages, so the term $ku^n = c$ enters in analysis only as a single constant, c. In different formulations, there will be different combinations of factors that together would be estimated as a single constant from data. I will use c to denote such constants, although the particular aggregate of factors subsumed by c may change from case to case.

With these assumptions, one may begin an analysis of dosage, d, and duration, τ, with an expression such as

$$I(\tau) \approx c(1 + bd)^r \tau^{n-1} \qquad (9.1)$$

or a suitably modified equation to match the particular problem.

If, as often assumed, moderate to large doses significantly increase transitions, then bd is much larger than one, and the transition rate becomes $u(1 + bd) \approx ubd$. Incidence is then

$$I(\tau) \approx cd^r \tau^{n-1} \qquad (9.2)$$

with incidence rising as the rth power of dose, d^r and the $n - 1$st power of duration, τ^{n-1}. Here, $c = ku^n b^r$, representing a single constant that may be varied or estimated from data.

Sometimes it is useful to study cumulative incidence, the summing up (integration) of incidence rates over the duration of exposure. This leads to the simple expression for cumulative incidence

$$CI(\tau) \approx cd^r \tau^n. \qquad (9.3)$$

Here, c differs from above but remains an arbitrary constant to vary or estimate from data.

In most empirical studies, incidence rises with a much lower power of dose than duration, $r < n$. This fact has led most authors to suggest that carcinogens typically affect only a subset of the transitions. For example, if one estimates $r = 2$ and $n = 6$, then one could interpret those results by concluding that the carcinogen affects two of the six transitions.

Later, I will suggest that this classic formulation of the theory may be misleading. In particular, the observation that the exponent on dosage is usually less than the exponent on duration does not necessarily imply that the carcinogen affects only a small number of transitions. However, the classic puzzle for the different responses to dosage and duration arises from the theory outlined here, so I use that theory as my starting point.

Cigarette Smoking

The classic study of cigarette smoking among British doctors estimated annual lung cancer incidence in the age range 40–79 as $I(\tau) \approx c(1+d/6)^2\tau^{4.5}$, where c is a constant, d is dosage measured as cigarettes per day, and $\tau = t - t_0$ is duration of smoking with t as age and $t_0 = 22.5$ as estimated age at which smoking starts (Doll and Peto 1978). If we use the expression for incidence in Eq. (9.1), then the estimate by Doll and Peto (1978) corresponds to $r = 2$ and $n = 5.5$.

Figure 9.1 shows the dose-response relationship for cigarette smoking, in which Doll and Peto (1978) fit a quadratic response curve. Subsequent authors have reiterated that lung cancer incidence increases with the first or second power of the number of cigarettes smoked per day (Zeise et al. 1987; Whittemore 1988; Freedman and Navidi 1989; Moolgavkar et al. 1989).

Carcinogen Applied to Laboratory Rats

Peto et al. (1991) presented a large dose-response experiment in which they applied the carcinogen N-nitrosodiethylamine (NDEA) to laboratory rats. I summarized the details of this experiment and other laboratory studies in Section 2.5. Here, I repeat the main conclusions.

Peto et al. (1991) measured, for each dosage level, the median duration of carcinogen exposure required to cause a tumor. Suppose we fit an empirical relation for the cumulative incidence rate, CI, which is the

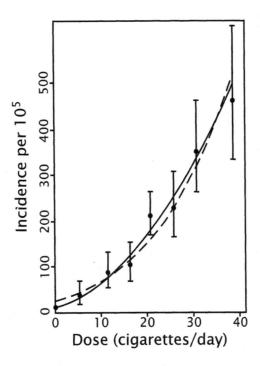

Figure 9.1   Dose-response for cigarette smoking, standardized for age. The filled circle and error bars mark the mean and 90% confidence interval at various dosages. The solid line shows the quadratic fit given by Doll and Peto (1978) with incidence per 10^5 equal to $9.36(1 + d/6)^2$. The dashed curve shows my calculation in which a nearly equivalent fit for incidence per 10^5 individuals can be obtained with a higher power of dose, in this case $25(1 + d/46)^5$. Redrawn from Figure 1 of Doll and Peto (1978).

total incidence over the duration of exposure (see Background above and Section 7.5). In empirical studies of dose-response, one typically observes that CI increases approximately with the rth power of dose and the nth power of duration, $CI(\tau) \approx cd^r t^n$. Then for the fixed level of cumulative incidence that occurs at the median duration to tumor development, $\tau = m$, we have $CI(m)/c \approx d^r m^n$. Taking the logarithm of both sides and solving for $\log(m)$ yields

$$\log(m) \approx (1/n)\log(k) - (r/n)\log(d), \tag{9.4}$$

where $k = CI(m)/c$ is a constant estimated from data. This equation is the expression of the classical Druckrey formula that I presented in Eq. (2.4).

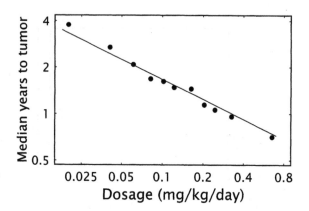

Figure 9.2   Esophageal tumor dose-response line. The circles show the observed durations of exposure required to cause one-half of the treatment group
to develop a tumor. Each median duration is matched to the dosage level for the
treatment group of rats. The line shows the excellent fit to the Druckrey formula
in Eq. (9.4) with $r = 3$, $n = 7$, $k = 0.036$, and a slope of $-r/n = -1/s = -1/2.33$.
Data from Peto et al. (1991).

Figure 9.2 shows that the results of Peto et al. (1991) fit closely to the
Druckrey relation with $n = 7$ and the slope $-r/n$ approximately $-3/7$,
leading to an estimate of $r = 3$. This analysis again shows that incidence
increases with a high power of duration and a relatively low power of
dose.

Zeise et al. (1987) reviewed many other examples of dose-response
relationships. In some cases, increasing dose causes a roughly linear
rise in incidence; in other cases, incidence rises with d^r, where d is dose
rate and $r > 1$, usually near 2; in yet other cases, incidence rises at a
rate lower than linear, with $r < 1$.

Perhaps only one pattern in dose-response studies recurs: the rise
in incidence with dose is usually lower than the rise in incidence with
duration of exposure, that is, $r < n$, as emphasized by Peto (1977).

ALTERNATIVE EXPLANATIONS

The observation that incidence rises more slowly with dosage than
with duration plays a key role in the history of carcinogenesis studies
and multistage theories. To give a sense of this history, I briefly list
some alternative explanations. I also comment on how well different
theories fit the observations: although fitting provides a weak mode of

discrimination, it does provide a good point of departure for figuring out how to construct informative comparative tests. I delay discussion of tests until later in this chapter.

CARCINOGENS AFFECT SOME STAGES BUT NOT OTHERS

Suppose, as discussed above, that a carcinogen increases the rate of transition between stages to $u(1 + bd)$, where d is dosage and b translates dose level into an increment in transition rate. If, for certain stages in progression, moderate to large doses significantly increase the transition rate, then bd is much larger than one, and the transition rate becomes $u(1 + bd) \approx ubd$. For other stages not much affected by the carcinogen, bd is small, and $u(1 + bd) \approx u$.

If a large increase in transition rate occurs for r of the stages, and the carcinogen has little effect on the other $n - r$ transitions, then as I showed in Eq. (9.3) above,

$$CI(\tau) \approx cd^r \tau^n,$$

with the cumulative incidence rate rising as the rth power of dose, d^r, and the nth power of duration, τ^n.

This explanation easily fits any case in which incidence increases exponentially with dosage and duration. However, the mathematics of curves provides no reason to believe that the number of steps affected by carcinogens can be inferred by measuring the empirical fit to the exponent on dosage.

THE MATHEMATICS OF CURVES: CARCINOGENS AFFECT ALL STAGES

Consider the most famous dose-response study: smoking among British doctors. Figure 9.1 shows the fit given by Doll and Peto (1978), in which the highest exponent of dose is two. From that fit, many authors have stated that lung cancer depends on the second power of dose, and thus the carcinogens in cigarette smoke affect only two stages in lung cancer progression. Against that explanation, the dashed curve in Figure 9.1 illustrates my calculation that a nearly equivalent fit for incidence can be obtained with a higher power of dose, in this case proportional to $(1 + d/46)^5$.

The fact that one can fit a higher power of dose to those lung cancer data certainly does not mean that the carcinogens in cigarette smoke affect five stages of carcinogenesis rather than two. It does mean that the original fit to the second power of dose provides little evidence with regard to the number of stages affected.

In general, an expression in a lower power of dosage, d, will often fit the data about as well as an expression in a higher power of d over a moderate range of dosage (Zeise et al. 1987; Pierce and Vaeth 2003). In fitting data, one usually prefers the fit from the lower exponent because it is regarded as more parsimonious. However, when trying to infer biological mechanism, moderate distinctions between the goodness-of-fit of expressions that have various exponents on d do not provide strong evidence about the number of stages affected by a carcinogen.

In the remainder of this section, I present some examples and technical issues about dose-response curves for those readers who like to see the details (see also Pierce and Vaeth 2003). Suppose a carcinogen affects all n transitions equally. Then dosage raises incidence by $k(1 + bd)^n$, where k is an arbitrary constant, and bd is the incremental increase in transition rate caused by dose d and scaling factor b. The expression for dosage can be expanded into a series of terms with increasing powers of d as

$$k\,(1 + bd)^n = k \sum_{i=0}^{n} \binom{n}{i} (bd)^i.$$

As bd declines, those terms with smaller exponents on d increasingly dominate the contribution of dosage, and so it would appear in the data as if the exponent on dose was small.

The smoking data in Figure 9.1 provide a good example. In those data, the exponent on duration suggests that $n \approx 6$, that dosage varies over a range of about 0–40 cigarettes per day, and that incidence increases by a factor of about 50 over the range of dosage studied. Using those data to provide reasonable ranges for dosage and for the consequences on incidence, suppose that a carcinogen affects incidence by the expression $k(1 + bd)^r$, with $k = 1$, $b = 1/43.5$, and $r = 6$. The solid curve in Figure 9.3 shows the dose-response effect.

In an empirical study, we would attempt to estimate the solid dose-response curve in Figure 9.3 from the data. The difficulty arises from the fact that we can get a good fit for $r = 2$, and that the fit improves relatively little for higher values of r. The figure shows example curves for $r = 2$ and $r = 3$ that fit very closely to the true curve. By the common statistical methods, one would usually choose the fit with the lower

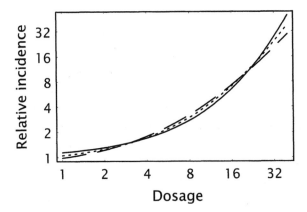

Figure 9.3 Lower power dose-response curves match higher power curves when dose and response vary over intermediate scales. Here, dosage varies over 1–40 and relative incidence in response to exposure varies over 1–50, matching the ranges in the smoking data of Figure 9.1. I scaled both axes logarithmically to analyze how a percentage increase in dose causes a particular percentage increase in relative incidence. All curves follow $k(1 + bd)^r$. In this theoretical example, the solid curve shows the true dose-response if the carcinogen affected $r = 6$ transitions, with $k = 1$ and $b = 1/43.5$. The long-dash curve shows the close fit to the true curve that can be obtained with $r = 2$ by choosing parameters that minimize the total squared deviations between the curves, $k = 0.77$ and $b = 1/7.7$. The short-dash curve shows that only a small improvement in fit can be obtained using a curve with $r = 3$, $k = 0.88$, and $b = 1/15.9$.

power of $r = 2$, noting that there is no statistical evidence that higher exponents fit the data significantly better.

DIMINISHING RISE IN CARCINOGENESIS AS DOSAGE INCREASES

Multistage analyses typically assume that, for each particular transition rate between stages, the carcinogen either has no effect or causes a linear rise in transition rate with increasing dose. Authors rarely discuss reasons for assuming a linear increase in transition rates with dose. A supporting argument might proceed as follows. Mutation rates often rise linearly with dose of a mutagen. If carcinogens act directly as mutagens, then carcinogens increase the rates of transition between stages in a linear way with dose.

Carcinogens may often act by processes other than direct mutagenesis. In particular, Cairns (1998) argued that carcinogens act mainly as mitogens, increasing the rate of cell division. Increased cell division does of course increase the accumulation of mutations, but does so differently from the mechanisms by which classical mutagens act. For example, the potentially mutagenic chemicals in cigarette smoke diffuse widely throughout the body, yet the carcinogenic effects concentrate disproportionately in the lungs. To explain this discrepancy in smokers between the distribution of chemical mutagens and the distribution of tumors, Cairns argued that the carcinogenic effects of smoke arise mostly from the irritation to the lung epithelia and the associated increase in cell division.

If carcinogens sometimes act primarily by increasing cell division, then we would need to know how mitogenic effects rise with dose. For example, doubling the number of cigarettes smoked might not double the rate at which epithelial stem cells divide to repair tissue damage. I do not know of data that measure the actual relation between mitogenesis and dose, but, plausibly, mitogenesis might rise with something like the square root of dose instead of increasing linearly with dose.

A diminishing increase in transition rates with dose would explain the observation that the exponent on dose is usually less than the exponent on duration. That observation is often expressed with the Druckrey equation that fits data from many studies of chemical carcinogenesis (Figures 2.11, 9.2). The Druckrey equation can be expressed as $k = d^r m^n$, where k is a constant, d is the dose level, and m is the median duration of carcinogen exposure to onset of a particular type of tumor. Usually, $r < n$, that is, the exponent on dose is less than the exponent on duration. Peto (1977) mentioned that, for carcinomas, r/n is often about $1/2$.

Now consider a simple multistage model with n stages and equal transition rates, u, between stages. Assume a carcinogen has the same effect on all stages, in which the transition rate is $uf(d)$, where $f(d)$ is a function of carcinogen dose, d. Then $k = [f(d)]^n m^n$, because the carcinogen has the same multiplicative effect on all n stages.

Suppose that the rise in transition rates diminishes with dose, for example, $f(d) = d^a$, with $a < 1$. Then the basic multistage model with all n transitions affected by a carcinogen leads to $k = d^{an} m^n$. If $a = r/n$, then we have the standard Druckrey relation, $k = d^r m^n$, which closely fits observations from many different experiments with $a = r/n \approx 1/2$.

Alternatively, we could use the more plausible expression $uf(d) = u(1 + bd^a)$, which leads to the multistage prediction $k = (1 + bd^a)^n m^n$.

This expression is, on a log-log scale, $\log(m) = k' - \log(1 + bd^a)$, and may often fit the data well. For example, in the large carcinogen study shown in Figure 9.2, if we use Peto's (1977) suggested value of $a = r/n = 1/2$, with fitted values for two parameters of $k' = 1.01$ and $b = 16$, we obtain a line that is almost exactly equivalent to the fit of the Druckrey formula shown in the figure.

The match of this diminishing effect theory to the observed relation in Figure 9.2 shows that the data fit equally well to a model in which the carcinogen affects only $r < n$ of the stages in progression or a model in which the effects of carcinogen dose rise at a diminishing rate with increasing dose.

Diminishing effects of carcinogens with dose readily explain the observation that $r < n$. At present, little information exists about how widespread such diminishing effects may be. Carcinogenic acceleration of mitogenesis provides a plausible mechanism by which diminishing effects may arise, but additional mechanisms probably occur.

HETEROGENEITY

Individuals vary in their susceptibility to carcinogens. Heterogeneity in susceptibility arises from both genetic and environmental factors. Lutz (1999) suggested that heterogeneity may tend to linearize the dose-response curve, that is, to reduce the exponent on dosage in such curves. Lutz based his argument on a graph that illustrated how the aggregate dose-response curve may form when summed over individuals with different susceptibilities. To evaluate this idea, I describe a few specific quantitative models. These models suggest that heterogeneity can influence the dose-response curves, but heterogeneity does not provide a convincing explanation for the widely observed low exponent on dose.

Consider the following rough calculation to illustrate the effect of heterogeneity on the dose-response curve. Suppose a carcinogen affects the relative risk of cancer, S. Let S depend on bd, where d is the dose, and b is a factor that scales the effect of dose on relative risk.

Heterogeneity in individual susceptibility enters the analysis through individual variability in b, the scaling factor that translates dose into an increment in transition rate between stages of progression. We need the value of S averaged over the different individual susceptibilities in the population. Let the probability distribution for the values of b among individuals be $f(b)$. The value of S for each level of susceptibility, b,

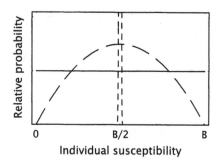

Figure 9.4 Distribution of individual susceptibility to carcinogens. For each individual, the consequence of carcinogen dose d scales with bd, where b is the individual's susceptibility to the carcinogen. This example uses the beta distribution to describe variation in individual susceptibility. The susceptibility values, b, range from 0 to a maximum of B. Two parameters, α and β, control the shape of the beta distribution. Here, I assume $\alpha = \beta$, so that all distributions have a symmetrical shape with mean $B/2$. The solid curve shows $\alpha = \beta = 1$; the long-dash curve shows $\alpha = \beta = 2$, and the short-dash curve shows $\alpha = \beta = 10,000$.

must be weighted by the various probabilities of different values of b. The average value of S over the different values of b is

$$S^* = \int Sf\,(b)\,\mathrm{d}b, \tag{9.5}$$

in which the distribution $f(b)$ describes the level of heterogeneity, and S is a function of b.

The slope of the dose-response curve on a log-log scale provides the empirical estimate for r, the exponent on dosage. The observed dose-response curve is S^*, so the log-log slope is

$$r = \frac{\mathrm{d}\log(S^*)}{\mathrm{d}\log(d)} = \frac{\mathrm{d}S^*}{\mathrm{d}d}\frac{d}{S^*}. \tag{9.6}$$

How does heterogeneity in individual susceptibility affect the shape of the dose-response curve? To study particular examples, we first need assumptions about the form of heterogeneity described by the distribution $f(b)$. Figure 9.4 shows three probability curves for heterogeneity, ranging from wide variation (solid line) to essentially no heterogeneity (tall, short-dashed curve).

Next, we need to assume particular shapes for the dose-response curve for a fixed level of susceptibility, that is, a fixed value of b. Figure 9.5 shows various examples. In the left panel, all the curves have

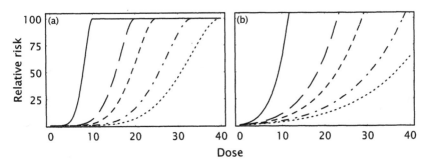

Figure 9.5 Relative risk, S, in response to dose, d. The plots show dose varying from 0 to 40, to illustrate roughly the range of dosage in number of cigarettes per day. However, the consequences of dose always depend only on bd, where b scales the dose into the actual effect. So the absolute dosage level does not matter, but the size of the interval does. (a) In this function, risk saturates to a maximum level, $S_m = 100$, at high dose for $bd > 1$, with $S = 1 + (S_m - 1)(bd)^n(n + 1 - nbd)$ for $bd < 1$. For all curves, $n = 6$. The curves, from left to right, show values of $b = 0.1, 0.05, 0.04, 0.03, 0.025$. (b) In this function, $S = (1 + bd)^n$, with all parameters as in panel (a).

a saturating response to high dose, above which relative risk no longer increases. In the right panel, risk continues to accelerate with increasing dose.

Figure 9.6 illustrates how heterogeneity affects the aggregate dose-response pattern in the population. In panel (a), the short-dash curve shows the dose-response pattern when there is essentially no heterogeneity. Increasing heterogeneity alters the shape of the dose-response curve, illustrated by the long-dash and solid curves of panel (a).

Figure 9.6b shows the log-log slopes of the aggregate dose-response curves, obtained by calculating the slopes of the curves in the panel above. These slopes provide the standard estimates for r, the exponent on dose in the dose-response relationship.

Figure 9.7 shows the same calculations, but for a base response curve that does not saturate at higher doses. In this case, heterogeneity always increases the slope of the dose-response curve.

The consequences of heterogeneity follow general rules. When the base curve rises at an increasing rate, then heterogeneity causes an increase in value because, at each point, the average of higher and lower doses is greater than the value at that point. By contrast, when the base curve rises at a decreasing rate, then heterogeneity causes a decrease in

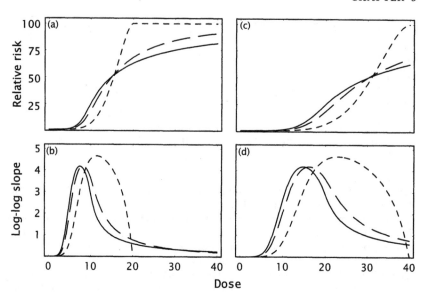

Figure 9.6   Consequences of heterogeneity in individual susceptibility on carcinogen dose-response curves. All curves derive from the response function
shown in Figure 9.5a: in panels (a) and (b), the average value of susceptibility
is $\bar{b} = 0.05$; in panels (c) and (d), the average is $\bar{b} = 0.025$. Panels (a) and (c)
show the dose-response curves when averaged over heterogeneity in susceptibility, calculated from Eq. (9.5). The three curves in each panel correspond to
the three distributions of susceptibility, b, in Figure 9.4. Panels (b) and (d) show
the corresponding log-log slopes of the dose-response curves, calculated from
Eq. (9.6).

value because, at each point, the average of higher and lower doses is
less than the value at that point.

In summary, large increases in heterogeneity usually cause minor
changes in the dose-response patterns. Those changes alter the details
of the dose-response relationship in interesting ways, but probably do
not explain the different effects of dosage and duration on incidence.

CLONAL EXPANSION

Precancerous stages in progression may proliferate by clonal expansion. The expanding clone of cells carries somatic mutations or other
heritable changes. I described the theory of clonal expansion in Section 6.5.

Clonal expansion could explain the different observed exponents on
dosage and duration. Suppose, for example, that cancer requires only

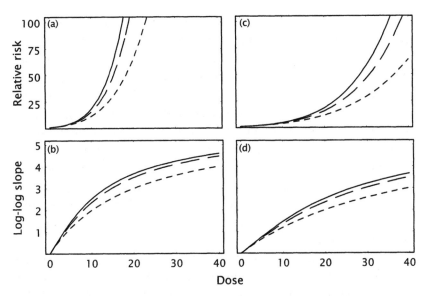

Figure 9.7 Consequences of heterogeneity in individual susceptibility on carcinogen dose-response curves. All curves derive from the response function shown in Figure 9.5b. Other assumptions match those described in Figure 9.6.

two rate-limiting transitions. The first transition causes the affected cell to expand clonally. As the number of cells in the clone increases, the rate of transition to the second stage rises because of the greater number of target cells available. In a carcinogen exposure study, incidence would rise with an increasing exponent on duration because the target population of cells for the final transforming step would increase with time.

A two-stage model could fit a variety of exponents for duration of smoking (Gaffney and Altshuler 1988; Moolgavkar et al. 1989), including the exponent of $n - 1 \approx 4.5$ reported by Doll and Peto (1978). The two-stage model could also fit the observed exponent on dosage of about two, because in a two-stage model the carcinogenic effects of smoking may influence two independent transformations.

Although the two-stage model cannot be ruled out, we do not know the exact nature of cancer progression and the rate-limiting steps that determine progression dynamics. I tend to favor other models for four reasons.

First, the ability of two-stage models to fit the data provides relatively little insight: with enough parameters and a mathematically flexible formulation, a model can be molded to a wide variety of data. Second, qualitative genetic evidence points to several rate-limiting steps in most adult-onset cancers (Chapter 3), although those data are not conclusive. Third, to explain the high observed exponents on age or duration, one must typically assume that clonal expansion is slow and steady over many years; bursts of clonal expansion over shorter periods do not match the observations so easily. Fourth, clonal expansion is more difficult to test experimentally than models that emphasize simple genetic or epigenetic changes to cells, because genomic changes can be manipulated and compared between treatments more easily than properties of clonal expansion.

The two-stage model may be limited and difficult to test. However, aspects of clonal expansion in multistage progression may play an important role in the patterns of incidence (Luebeck and Moolgavkar 2002). To move ahead, this idea requires useful comparative hypotheses that predict different outcomes based on measurable differences in the dynamics of clonal expansion.

SUMMARY

Several theories fit the observed relatively low exponent on dosage and high exponent on duration. But a close fit by itself provides little evidence to distinguish one theory from another. Rather, one should use the alternative theories and fits to the data as a first step toward developing biologically plausible hypotheses and their quantitative consequences. Once those theories are understood, one can then try to formulate comparative tests that discriminate between the alternatives. I turn to potential comparative tests after I discuss a related topic in chemical carcinogenesis.

9.2 Cessation of Carcinogen Exposure

Lung cancer incidence of continuing smokers increases with approximately the fourth or fifth power of the duration of smoking (Doll and Peto 1978). By contrast, incidence among those who quit remains relatively flat after the age of cessation (Doll 1971; Peto 1977; Halpern et al. 1993).

In 1977, Richard Peto (1977) stated that the approximately constant incidence rate after smoking ceases "is one of the strongest, and hence most useful, observational restrictions on the formulation of multistage models for lung cancer." Peto argued that, in any model, the observed constancy in incidence after smoking has stopped "suggests that smoking cannot possibly be acting on the final stage" of cancer progression. There could, for example, be a particular gene or pathway that acts as a final barrier in progression and resists the carcinogenic effects of cigarette smoke.

In 2001, Julian Peto (2001) reiterated Richard's argument: "The rapid increase in the lung cancer incidence rate among continuing smokers ceases when they stop smoking, the rate remaining roughly constant for many years in ex-smokers (Halpern et al. 1993). The fact that the rate does not fall abruptly when smoking stops indicates that the mysterious final event that triggers the clonal expansion of a fully malignant bronchial cell is unaffected by smoking, suggesting a mechanism involving signaling rather than mutagenesis."

In this section, I discuss which stages of progression may be affected by the carcinogens in cigarette smoke. I begin by summarizing observations on how cancer incidence changes after the cessation of carcinogen exposure. I then consider two alternative explanations. First, the carcinogen may affect only a subset of stages in cancer progression; the particular stages affected determine how patterns of incidence change after cessation. Second, the carcinogen may affect all stages of progression; the different precancerous stages at which individuals cease exposure determine how patterns of incidence change after cessation. Both models fit the data reasonably well.

As we have seen often, fitting by itself does not strongly distinguish between competing hypotheses. I therefore introduce some comparative approaches that may provide a better way to test alternatives.

OBSERVATIONS

Figure 9.8a shows the flattening of the incidence curve upon cessation of smoking from data collected in the Cancer Prevention Study II of the American Cancer Society (Stellman et al. 1988). This figure summarizes data for 117,455 men who never smoked, 91,994 current smokers, and 136,072 former smokers. The top curve represents lifetime smokers

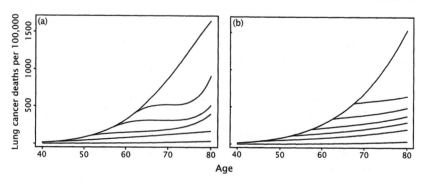

Figure 9.8   Reduction in relative risk of lung cancer between men who continued to smoke and those who quit at different ages. (a) Summary of data from
Figure 1 of Halpern et al. (1993). The top curve shows those who continued
to smoke. The lower curves show those who quit at different ages, the age of
quitting marked by the intersection of a lower curve with the top curve. The
bottom curve describes those who never smoked. Sample sizes given in the
text. (b) Model fit to the data in which smoke carcinogens affect equally all
stages in progression. The subsection *All Stages Affected* describes the details
of the model.

who never quit. The four curves below it represent individuals who quit
at different ages; the age at which smoking ceased coincides with the
intersection of each curve with the top curve for lifetime smoking. The
bottom curve shows incidence among those who never smoked.

Figure 9.9a presents data from a cessation of smoking study in the UK
(Peto et al. 2000). That study analyzed cumulative risk rather than incidence rate. Cumulative risk measures the lifetime probability of death
from lung cancer at each age if no other causes of death were to occur.
A flat incidence rate translates into a linear increase in cumulative risk
with age. The plot shows that cessation of smoking reduces the upslope
in cumulative risk, somewhere between linear (flat incidence) and the accelerating curve for those who continue to smoke. Thus, the pattern in
Figure 9.9a matches the pattern in Figure 9.8a: an initial flattening of the
incidence rate after cessation of smoking followed by a relatively slow
rise later in life.

Other studies report data on cessation of carcinogen exposure (reviewed by Day and Brown 1980; Freedman and Navidi 1989; Pierce and
Vaeth 2003). I focus only on the smoking data, because those studies
have the largest samples and have been discussed most extensively. I
emphasize how to develop and test hypotheses rather than argue for

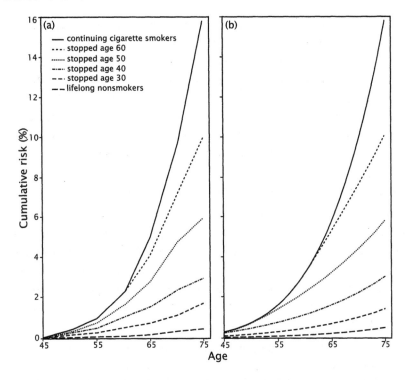

Figure 9.9   Reduction in relative risk of lung cancer between men who continued to smoke and those who quit at different ages. (a) Redrawn from Figure 3 of
Peto et al. (2000). Samples for this case-control design include 1465 case-control
pairs in a 1950 study combined with 982 cases plus 3185 controls in a 1990
study. (b) Model fit to the data in which smoke carcinogens affect equally all
stages in progression. The subsection *All Stages Affected* describes the details
of the model.

a comprehensive explanation to cover all of the available data. In my
opinion, the existing studies do not provide enough evidence to decide
between competing hypotheses. Instead, the smoking data define the
challenge for future studies.

ALTERNATIVE EXPLANATIONS

All theories must account for two observations. First, the relative risk
of lung cancer decreases in those who quit compared with those who
continue to smoke (Figures 9.8 and 9.9). Second, the rise in incidence
with smoking fits an increase in incidence with roughly the second power
of number of cigarettes smoked per day (dose).

I discuss two alternative formulations. First, most prior explanations fit the observations by positing that carcinogens in smoke affect only one or two stages in progression, leaving the other stages mostly unaffected.

Second, I show that the standard multistage model of progression also fits the observations very well. Previous authors rejected that standard model because they used the common approximation for incidence given by Armitage and Doll (1954), which in fact does not apply well to the problem of carcinogen exposure followed by cessation.

CARCINOGENS AFFECT SOME STAGES BUT NOT OTHERS

This idea was stated most clearly and perhaps originally by Armitage in the published discussion following Doll (1971). I quote from Armitage at length, because his words set the line of thinking that has dominated the subject. Note that, at the time, the dose-response curve was thought to be linear. Later work suggested that the response may in fact fit a curve that rises with the square of dose (Doll and Peto 1978). Here is what Armitage said:

> The dose-response relationship seems to be linear, which suggests that the carcinogen affects the rate of occurrence of critical events at one stage, and one only, in the induction period. (If it affected two stages, one might have expected a quadratic relationship, and so on.) Does this crucial event happen early or late in the induction period? For example, in a six-stage process, are we thinking of an early stage, the first or second, or a late stage, the fifth or sixth?
>
> The evidence here seems to conflict. One argument would suggest that a very early stage is involved. I am thinking of the delay of a generation or so between the increase in smoking in men around the First World War, and the rise in lung cancer mortality rates which was so marked 20 or 30 years later; and similarly the increase in cigarette smoking among women about the time of the Second World War, and the rise in lung cancer rates for females which has become so noticeable in the last few years. This long delay is what one would expect if a very early part of the process were involved rather than a very recent one.
>
> On the other hand, the halt in the rise in risk quite soon after smoking stops suggests that a late stage is involved. Professor Doll's very ingenious treatment of the data on ex-smokers, in Tables 13 and 14, confirms the latter view. In a multi-stage process, if the first stage were involved, the rate after stopping smoking would continue to rise in the same way as for continuing smokers. If,

on the other hand, the last stage were affected, one would expect the rate to drop immediately to the rate for nonsmokers. What seems to happen is a stabilization at the current rate until it is caught up by the rate for nonsmokers. That is precisely what one would expect if the next to last stage in a multi-stage process were affected.

I should be interested to know whether Professor Doll has considered this anomaly and can resolve it. Is it, for example, conceivable that two stages in a multi-stage process are affected ...?

Exactly how does incidence change when a carcinogen affects only one of n stages? Whittemore (1977) and Day and Brown (1980) presented approximate theoretical analyses. However, those approximations can be rather far off from the actual theoretical values. I prefer exact calculations as shown in the example of Figure 9.10. I describe in detail the results in Figure 9.10, because this particular model played an important role in the history of carcinogen studies. The model also provides general insight into multistage progression.

In Figure 9.10a,b, I used a basic n stage model in which a carcinogen increases the rate of the ith transition between stage i and stage $i + 1$. For example, if $i = 0$, then the carcinogen affects only the first transition between the baseline stage 0 and the first precancerous stage 1; if $i = n - 2$, then the carcinogen affects only the penultimate transition between stage $n - 2$ and stage $n - 1$. The model in Figure 9.10 has $n = 6$ stages. The legend shows the line types that describe the outcome when the carcinogen affects the ith transition.

In Figure 9.10a,b, the carcinogen is applied only between age 0 and age 60, after which carcinogen application ceases. If the carcinogen affects one of the first three transitions, shown in Figure 9.10a, then incidence follows closely the curve that would result if the carcinogen was applied throughout life, from age 0 to age 80. With acceleration of an early stage, cessation has little effect on incidence because anyone who ultimately progresses to cancer has already passed the early stages by age 60.

Figure 9.10b shows the strong effect that cessation has on incidence when a carcinogen is applied from age 0 to age 60 and influences a later stage in progression. If the carcinogen affects the last transition, $i = 5$, then during carcinogen application, anyone who progresses to the fifth stage is almost immediately transformed into the final cancerous stage. Thus, the curve for $i = 5$ up to age 60 shows the incidence pattern for a five-stage model: the six stages of progression minus one stage that is not rate limiting in the presence of the carcinogen. After cessation, progression follows the full six rate-limiting stages, and so incidence instantly drops to the rate for a six-stage model.

If the carcinogen affects only the penultimate transition, $i = 4$, then during carcinogen application, individuals move very rapidly from stage 4 to stage 5, where they await the final transforming event at the normal, background rate. By essentially skipping a stage during carcinogen application, the incidence follows a five-stage model. After cessation, almost all new cancers arise from the pool of individuals in stage 5 who await the final transition. When transformation occurs by a single random event, the incidence rate remains flat over time. The final event is as likely to happen this year as next year or a later year. If the carcinogen affected only the third transition, $i = 3$, then after cessation most cancers would arise in the pool of individuals that require two further steps, causing incidence to increase only slowly with time as in a model with only two stages.

In Figure 9.10c,d, the carcinogen is applied only between age 25 and age 80. The carcinogen has relatively little effect when it increases the earliest transition, $i = 0$, because that transition has already occurred by age 25 in many of the individuals who ultimately progress to cancer. For the next transition, $i = 1$, fewer would have passed that step by age 25, and so more will be affected by the carcinogen. For the later steps, almost no one would have passed those steps by age 25, and so the carcinogen increases incidence equally for all of the later transitions.

In Figure 9.10e,f, the carcinogen is applied only between age 25 and age 60, after which carcinogen application ceases. This case matches the problem of cessation smoking, with onset of smoking in the first third of life and cessation in the last third of life. The patterns can be understood from the previous cases. If smoking affects only an early stage, then the earlier the stage, the less the effect, because the earliest stages are more likely to have been passed already before the onset of smoking and the acceleration of that stage. If smoking affects only a later transition, i, then after cessation, the pool of individuals most susceptible has $n - i$ steps remaining; if smoking affects the final transition, no excess pool of susceptibles exists, and incidence reverts to the background rate.

The first theoretical studies of smoking cessation considered models in which smoking affected only one stage (Whittemore 1977; Day and

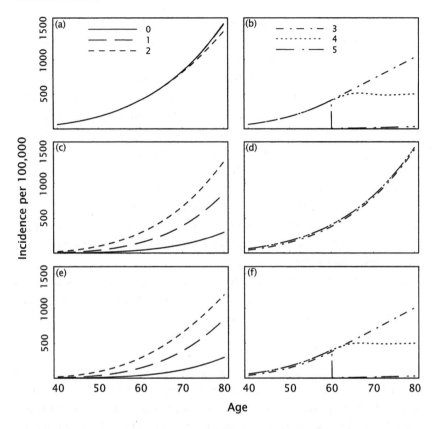

Figure 9.10 Theoretical incidence curves in response to carcinogen application followed by cessation. The carcinogen affects only a single transition in a model with $n = 6$ steps. The legend shows the curve type for each of the $i = 0,\ldots,5$ transitions, in which the carcinogen affects only the ith transition. (a and b) Carcinogen applied from age 0 to age 60. (c and d) Carcinogen applied from age 25 to age 80. (e and f) Carcinogen applied from age 25 to age 60. I calculated the curves by numerical evaluation of the complete progression dynamics as described in earlier chapters. I used the following assumptions: the number of lineages per individual, $L = 10^8$; the transition rate for steps not affected by the carcinogen, $u = 7.24 \times 10^{-4}$; and the transition rate for the single step affected by the carcinogen during those ages of exposure, $u(1 + d)$, where $d = 70$.

Brown 1980). The analyses I just presented improve the accuracy of such models over previous studies, but the main points hold from earlier work. After that early work, two observations affected subsequent analyses of smoking cessation. First, none of the curves in Figure 9.10

fit closely to data such as in Figure 9.8. Second, later studies of dose-response favored a quadratic fit to the data, leading many to suppose that smoking affects two stages in progression.

One can see from Figure 9.10e,f that a combination of the earliest transition, $i = 0$, and the penultimate transition, $i = n - 2 = 4$, provides the shapes needed to fit the data in Figure 9.8, and with two transitions affected, the overall incidence would be higher. Various authors fit the data in this way, sometimes weighting the role of those two stages differently (Day and Brown 1980; Brown and Chu 1987; Whittemore 1988).

Those fitted models based on two affected stages match the data reasonably well for both dose-response and incidence. In particular, one can easily explain the flattening of the incidence curves upon cessation by the penultimate transition and the later rise in incidence several years after cessation by the earliest transition.

The data and matching models tell a pleasing empirical and logical story. However, other plausible models also fit nicely to the data. The next section provides an example.

ALL STAGES AFFECTED

Armitage's quote shows that the linear or perhaps quadratic dose-response curve motivated the initial models in which smoke carcinogens affect only one or two stages of progression. Those assumptions about number of stages affected may over-interpret the data: one cannot draw firm biological conclusions about a molecular mechanism from a fitted exponent of a dose-response curve. In addition, the mathematical analyses of progression have in the past typically used approximations; those approximations do not capture key aspects of incidence curves and dose-response curves.

I decided to analyze how well the standard model of multistage progression fits the data, in which the carcinogens affect equally all n stages. I first fit the data in Figure 9.8a, giving the fitted curves shown in Figure 9.8b. To obtain those fitted curves, I began with the basic multistage model described earlier in the theory chapters. I took the following parameters as given based on previous studies or on common assumptions: the number of stages, $n = 6$; the number of independent cell lineages at risk, $L = 10^8$; the age at which smoking starts, 25 years; and the maximum age of the analysis, 80 years. Those parameters were not fit to the data but instead derived from extrinsic considerations.

I then used the following crude procedure to fit the model to the data. I set the cumulative lifetime risk of lung cancer for nonsmokers to 0.005 to match the lowest curve in Figure 9.8, which shows data for nonsmokers. I then fit the transition rate between stages per year, u, needed to match that nonsmoker incidence curve, resulting in the estimate $u = 7.24 \times 10^{-4}$. Given this value for the baseline transition rate, I next assumed that during exposure to smoke carcinogens, all transitions between stages rise to $u(1 + bd)$, where d is dose, and bd is the increase in the transition rate caused by carcinogens. The value of b sets a proportionality constant for the effect of a given dose; without loss of generality, I used $b = 1$, because all calculations depend only on the product bd and not on the separate values of the two parameters.

I estimated the value of $d = 1.187$ to match the top curve, in order to obtain a lifetime cumulative risk for continuing smokers of 0.158. Finally, I assumed that, upon cessation of smoking, carcinogenic effects decay with a half-life of 5 years; this assumption prevents an unrealistic instantaneous decline in incidence immediately upon cessation.

This fitting procedure required estimation of only two parameters, u and d. The other values came from prior studies or common assumptions. The fit shown in Figure 9.8b provides a reasonable qualitative match to the observed patterns in Figure 9.8a; some deviation occurs at age 80—a few observations at this point cause some of the incidence curves to rise late in life. Better fit could be obtained by optimizing the fit procedure and by using additional parameters. But my point is simply that the basic multistage model gives a nice match to the data without the need for any special adjustment or refined fitting.

Originally, Armitage, Peto, and others rejected a model in which carcinogens affect all stages because the estimated exponent of the dose-response curve is between one and two. Does the model I used, with all stages affected, also match that observed dose-response relation?

To test the model fit to the observed dose-response curve, I focused on the estimated value of d, which in the standard models is proportional to dose. At the maximum age measured, in this case 80 years, I varied the cumulative lifetime risk for continuing smokers between the value for nonsmokers, 0.005, and the approximate observed value for lifetime smokers of 0.158. For each cumulative risk value (the response), I fit the d value (the dose) needed to match the cumulative risk. I then calculated the log-log slope of the dose-response curve, which turned out to be

1.84. Thus, the model provides a good match to the observed exponent on the dose-response relation. The earlier section, *The Mathematics of Curves,* and Figure 9.3 explain why a model with $n = 6$ steps can give an approximately quadratic dose-response curve.

I repeated the same fitting procedure for the data in Figure 9.9a. In those data, the maximum observed age is 75; otherwise, I used the same background assumptions as in the previous case. The shift in maximum observed age altered the two fitted parameters: $u = 7.72 \times 10^{-4}$ and $d = 1.225$. The model provides a close fit to the data (Figure 9.9b). The log-log slope of the dose-response curve is 1.84, as in the previous case.

In summary, a model with all stages affected fits the data reasonably well. The data do not provide any easy way to distinguish between this model, with all stages affected, and the earlier models in which the carcinogens affect only one or two stages. Perhaps the most striking difference arises in the carcinogenic increase in transition rate that one must assume: when the carcinogen affects all stages, the increase, d, is about 1.2, or 120 percent. This small increase in transition would be consistent with a moderate and continuous increase in cell division: the mitogenic effect perhaps caused by irritation. By contrast, when the carcinogen affects only one stage, the required increase in transition rate, d, may be around 70, and for two stages, d is probably around 8–10. Those large increases in transition seem too high for a purely mitogenic effect, and would therefore point to a significant role of direct mutagenesis in increasing progression.

Fitting models cannot decide between mitogenic and mutagenic hypotheses. In the next section, I discuss how to use the quantitative models as tools to formulate testable hypotheses.

9.3 Mechanistic Hypotheses and Comparative Tests

Two observations set the puzzle. First, cancer incidence rises more rapidly with duration of exposure than with dosage. In terms of lung cancer, incidence rises more rapidly with number of years of smoking than with number of cigarettes smoked per year. Second, lung cancer incidence remains approximately constant after cessation of smoking but rises in continuing smokers.

Traditional explanations suggest that carcinogens affect only a subset of stages in progression. Such specificity in carcinogenic effects would often lead to incidence patterns that fit the observations.

I discussed in the previous section how an alternative model in which carcinogens affect all stages also fits the observations. The fact that the observations can be fit by a model in which all stages are affected does not argue against the traditional explanation in which only a few stages are affected. Rather, the proper inference is that we need to be cautious about drawing firm conclusions about mechanism solely from models fit to age-incidence curves.

Further progress requires testing alternative hypotheses about the link between, on the one hand, how carcinogens affect the mechanisms of progression dynamics and, on the other hand, how perturbations of progression dynamics cause shifts in the age-onset curves. I focus on shifts in age-onset curves because carcinogenic perturbations are important only to the extent that they cause changes in incidence patterns.

In this section, I present alternative mechanistic hypotheses about how carcinogenic perturbation affects progression dynamics. I also consider the sorts of comparative tests that could distinguish between alternative mechanistic hypotheses.

BACKGROUND

Tumors arise when cell lineages evolve ways around the normal limits on tissue growth. Because tumors develop through evolutionary processes, we can classify the mechanisms of carcinogen action by the particular evolutionary processes that they affect.

Variation and selection comprise the most important evolutionary processes. For variation, I consider carcinogenic effects that act directly by mutagenesis, defined broadly to include karyotypic and epigenetic change. The different types of heritable change cause different spectra of variation and act at different rates. For selection, I divide mechanisms into three classes: mitogens directly increase cellular birth rate, anti-apoptotic agents directly reduce cellular death rate, and selective environment agents favor cell lineages predisposed to develop tumors. Those selective mechanisms may indirectly increase variation. For example, mitogens often increase mutation by raising the rates of DNA replication.

I do not use the common classification that divides the effects of carcinogens into initiation, promotion, and progression. That classification primarily arises from the tendency of certain agents, at certain doses, to have stronger effects when applied before or after other agents. Such patterns certainly exist and must, to some extent, be correlated with mechanism of action. Indeed, initiators do sometimes act as direct mutagens that cause particular mutations early in tumor formation, and promoters do often act as mitogens. But there are many exceptions with regard to the consistency of the patterns, and the connections to mechanism often remain vague and somewhat speculative (Iversen 1995).

My focus on variation and selection does not set a mutually exclusive alternative against the classical initiation-promotion-progression scheme. Instead, my emphasis on variation and selection simply puts the processes of tumor evolution ahead of the sometimes debatable patterns for the ordering of consequences under certain experimental conditions.

I place carcinogenic mechanism in the context of multistage progression, measured by shifts in age-onset curves. I therefore emphasize how certain mechanisms affect rate processes and the time course of tumor formation. For example: How does a carcinogenic agent affect the rate of transition between particular stages? How many stages does an agent affect? Does a particular agent have an effect only on tissues that have already progressed to a certain stage? Put concisely, the issues concern changes in rate, number of stages affected, and order of effects.

MUTAGENS: INCREASE HERITABLE VARIATION

I begin with background observations from the mouse skin model of chemical carcinogenesis (Slaga et al. 1996). I then interpret those observations in terms of hypotheses about rate, number, and order.

BACKGROUND

The first step in skin tumor development often appears to be a mutation to H-*ras* that causes an amino acid substitution at codon positions 12, 13, or 61 in the phosphate binding domain of the protein (Brown et al. 1990). Those substitutions can abrogate negative regulation of the Ras signal that stimulates cell division (Barbacid 1987).

Different carcinogens induce different spectra of mutation to H-*ras* isolated from papillomas or carcinomas of mouse skin. Table 9.1 shows

Table 9.1   Carcinogen-induced H-*ras* substitutions in mouse skin papillomas

Carcinogen*	Substitution (codon)	Frequency in papillomas
MNNG	$G^{35} \rightarrow A$ (12)	11/15
MNU	$G^{35} \rightarrow A$ (12)	5/12
DMBA	$A^{182} \rightarrow T$ (61)	45/48
MCA	$G^{182} \rightarrow T$ (61)	4/20
MCA	$G^{38} \rightarrow T$ (13)	4/20

* Abbreviations: MNNG, N-methyl-N'-nitro-N-nitrosoguanidine; MNU, methylnitrosourea; DMBA, 7,12-dimethylbenz[a]anthracene; MCA, 3-methylcholanthrene. Initial carcinogen treatment followed by repeated application of TPA, 12-O-tetradecanoyl-13-acetylphorbol. Data from Brown et al. (1990).

the most frequent DNA base substitutions in response to four different carcinogens, measured in papillomas that did not progress to carcinomas. In this case, the carcinogens were applied in one dose at the start of treatment (an initiator), and most likely acted as direct mutagens. The initial treatment with one of the mutagens listed in Table 9.1 was followed by repeated application of a mitogen, TPA.

The observed substitution spectrum in response to an initial carcinogen probably results from two processes. First, the initial carcinogen treatment causes a particular spectrum of genetic changes. That primary spectrum depends on the biochemical action of the carcinogen with respect to DNA damage and repair. Second, among the variation caused by those initial changes, only certain mutations become amplified to form papillomas. In this case, selection amplified those cells that carry changes to the Ras protein and abrogation of negative regulation of mitogenic signals.

I summarized results on H-*ras* mutation (Table 9.1) to emphasize that different carcinogens often cause different spectra of heritable variation. Several other studies report carcinogen-specific spectra of heritable change to DNA sequence, epigenetic marks, or karyotypic alterations (reviewed by Lawley 1994; Turker 2003).

Mutation of H-*ras* appears to be a common early step of skin carcinogenesis in both mice and humans (Brown et al. 1995). Two alternative hypotheses could explain why H-*ras* mutations arise early in experimental studies of chemical carcinogenesis in mice. First, the particular carcinogens may produce a mutational spectrum that favors H-*ras* variation and selection. Second, amplification of H-*ras* mutation may be a favored early step in skin carcinogenesis, so that early change in H-*ras* is not strongly dependent on the particular spectrum of heritable change caused by a direct mutagen.

How do chemical carcinogens affect different stages of progression? The stage at which *p53* mutations occur in skin carcinogenesis and the spectrum of mutations to that gene provide some clues (Brown et al. 1995; Frame et al. 1998). Burns et al. (1991) observed no *p53* mutations in benign papillomas, an early stage in carcinogenesis, whereas they found that 25% of later stage carcinomas had *p53* mutations. It may be that early *p53* mutations are actually selected against in skin carcinogenesis. In three different studies that applied an initial mutagen to mouse skin, heterozygote $p53^{+/-}$ mice had fewer papillomas than did wild-type $p53^{+/+}$ mice (Kemp et al. 1993; Greenhalgh et al. 1996; Jiang et al. 1999). Another study showed that $p53^{+/-}$ mice had a three-fold increase in progression of papillomas to carcinomas, demonstrating a causal role of *p53* mutation in later stages of carcinogenesis (Brown et al. 1995).

In three different chemical carcinogen treatments of mouse skin, the particular spectrum of *p53* mutations depended on the treatment. When an initial mutagen, DMBA, was followed by the mitogen, TPA, most *p53* changes were loss of function mutations, including frameshifts, deletions, and the introduction of stop codons. Repeated application of DMBA led to five carcinomas with one deletion and four transversion mutations in *p53*. Repeated application of the mutagen MNNG led to four carcinomas with G → A transitions in *p53* (Brown et al. 1995).

HYPOTHESES AND TESTS

I describe a series of hypotheses and tests to show how one might in principle connect particular mechanisms of carcinogen action to consequences for multistage carcinogenesis. Some of the tests may not be experimentally well posed or practical to do, but they should help to stimulate thought about how to develop new, more practical tests that provide information about mechanism.

Measuring a rate of transition directly is difficult, so I focus on the number of transitions and the order of effects.

Hypothesis for number of steps affected by a carcinogen.—A treatment affects only a subset of rate-limiting steps.

Test.—Apply a mutagen continuously. If all steps are affected equally, then untreated and treated animals should have approximately the same slope of the incidence curve (log-log acceleration, LLA), because they have the same number of rate-limiting steps. The treated animals should, however, have a higher intercept for their age-incidence curve, because their transitions happen at a faster rate. If some transitions are more sensitive than others, then the LLA of the incidence curve should decrease with increasing dose because, as dose rises, an increasing number of steps should change from rate limiting to not rate limiting. The fewer the number of rate-limiting steps, the lower the LLA.

Hypothesis for mechanism of initial carcinogen treatment.—The primary effect is mutation of the first rate-limiting step in multistage progression.

Test.—Compare age-onset curves in mice with wild-type H-*ras* and H-*ras* mutated in one of the carcinogenic codons, each mouse genotype either treated or not treated with a single dose of an initial carcinogen. To get enough tumors for comparison, the mice could have a cancer-predisposing genotypic background with changes distinct from the functional consequences of H-*ras* mutation. If the initial carcinogen treatment only has a tumorigenic effect through mutation of H-*ras* as the first rate-limiting step, then the untreated, wild-type mice would have to pass one more step than either of the other three treatments: mutated H-*ras* with or without initial carcinogen treatment and wild-type H-*ras* treated with an initial carcinogen. An additional rate-limiting step to pass should cause the slope of the incidence curve (LLA) to be one unit higher than in treatments that rapidly pass that step.

Hypothesis for order of stages affected.—Certain carcinogens affect only a particular transition in an ordered series of stages of progression.

Test.—Suppose carcinogen *A* is thought to affect primarily an early stage, such as H-*ras* mutation in skin tumors, and carcinogen *B* is thought to affect primarily a late stage, such as *p53* mutation in skin tumors. The following comparisons support the hypothesis. If *A* acts early and *B* acts late, then the difference in incidence between *A* early and *A* late is greater than the difference between *B* early and *B* late. If *A* acts early and *B* acts late, then the combination of *A* applied early and *B* applied late has a stronger effect than *B* applied early and *A* applied late.

Summary.—These tests emphasize treatments that apply chemical carcinogens to altered animal genotypes, with age-incidence curves measured as the outcome and interpreted in the light of quantitative predictions of multistage theory.

Mitogens: Increase Cellular Birth Rate

Increased cell division raises the rate of tumor formation (reviewed by Peto 1977; Cairns 1998). Higher rates of tumorigenesis occur in response to irritation, wound healing, and chemical mitogens.

I first describe three hypotheses to explain the association between mitogenesis and carcinogenesis. Ideally, I would follow with tests that clearly distinguish between alternative hypotheses. However, given the current level of technology, it is not easy to define practical experiments that connect biochemical changes caused by mitogens to consequences for rates of tumorigenesis. With that difficulty in mind, I finish by laying a foundation for how to formulate tests as understanding and technology continue to improve.

HYPOTHESES

Faster cell division balanced by increased cell death.—In this case, the number of cells does not increase because tissue regulation balances cell birth and death, but the mitogen increases cell division and turnover. The faster rate of DNA replication increases the rate at which mutations occur (Cunningham and Matthews 1995).

Normally asymmetrically dividing cell lineages divide symmetrically.—Epithelial stem cells sometimes divide asymmetrically. One daughter remains as a stem cell to provide for future renewal; the other daughter often initiates a rapidly dividing and short-lived lineage. Cairns (1975) suggested that in each asymmetric stem cell division, the stem lineage may retain the older DNA templates, with the younger copies segregating to the other daughter cell (supporting evidence in Merok et al. 2002; Potten et al. 2002; Armakolas and Klar 2006). If most mutations occur in the production of new DNA strands, then most mutations would segregate to the nonstem daughter lineage, and the stem lineage would accumulate fewer mutations per cell division. In addition, stem cells may be particularly prone to apoptosis in response to DNA damage, killing themselves rather than risking repair of damage (supporting evidence in Bach et al. 2000; Potten 1998).

If these processes reduce stem cell mutation rates, then carcinogens or other accidents that kill stem cells may have a large effect on the accumulation of mutations (Cairns 2002). In particular, lost stem cells must be replaced by normal, symmetric cell division with typical mutation rates that may be much higher than stem cell mutation rates. Thus, regeneration of stem cells following carcinogen exposure or during wound healing may cause increased mutation.

Clonal expansion of predisposed cell lineages.—Once a mutation occurs, a mitogen may stimulate clonal expansion. An expanding clone increases the number of target cells for the next transition (Muller 1951). This increase in transition rate between stages does not require a rise in mutation rate per cell division, only an increase in the number of cells available for progressing to the next stage.

TESTS

The mechanistic details of mitogenesis may be studied directly at the biochemical and cellular levels. However, I am particularly interested in the different ways in which mitogenesis shifts age-incidence curves. To study shifts in age incidence, one must analyze how mechanistic consequences of mitogenesis affect rates at which carcinogenic changes accumulate in cells.

The first two mechanistic hypotheses in the previous section focus on an increase in the mutation rate per cell; the third hypothesis focuses on an increase in the number of target cells susceptible for transition to the next stage. The two processes have different consequences for age-incidence curves.

Increase in mutation rate per cell.—In this case, the mitogen acts like a mutagen. The particular hypotheses and tests from the section on direct mutagenesis apply.

Increase in number of target cells for next transition.—More target cells cause a higher transition rate per unit time. The main difference from mutagenic agents arises from the time course over which the mutation rate increases. When a chemical agent causes an increased rate of mutation per cell, the rise in the mutation rate most likely occurs over a short period of time. By contrast, an increase in the number of target cells may happen slowly as a predisposed clone expands, causing a slow rise in the transition rate to the following stage.

In the theory chapters, I demonstrated a clear difference in how age-incidence curves shift in response to a change in transition rate. A quick rise in a particular transition abrogates a rate-limiting step and reduces the slope of the age-incidence curve. In an idealized model, each abrogation of a rate-limiting step reduces the slope by one unit. By contrast, a slow rise in a transition rate causes a slow rise in the slope of the age-incidence curve. Multiple rounds of slow clonal expansion can lead to high age-incidence slopes. (See Section 6.5, which describes the theory of clonal expansion.)

Increasing the dosage of a mitogen may cause more rapid clonal expansion. The theory predicts that the increase in the rate of clonal expansion causes a steeper rise in the slope of the incidence curve over a shorter period of time. If the rate of clonal expansion is not too fast, then longer duration of exposure to a mitogen may cause a sequence of clonal expansions as one transition follows another, leading to a steep rise in the slope of the incidence curve. At high doses and rapid rates of clonal expansion, transitions may occur so rapidly that the rate-limiting effects of a stage may be abrogated, causing a drop in the slope of the incidence curve.

Anti-Apoptotic Agents: Decrease Cellular Death Rate

Anti-apoptosis may act in at least two different ways. First, blocking cell death may allow mutations to accumulate at a faster rate, because apoptosis is an important mechanism for purging damaged cells. Second, absence of cell death may cause clonal expansion, with an increase in the number of target cells for the next transition.

I discussed in the previous sections some of the ways in which to study increased mutation rate per cell versus increased target size in an expanding clone of cells. It may be possible to complement those approaches by study of genotypes with loss of apoptotic function.

Selective Environment: Favors Predisposed Cell Lineages

The previous sections discussed carcinogens that directly cause mutations or directly affect cellular birth or death. This section focuses on carcinogens that change the competitive hierarchy between genetically or epigenetically variable cell lineages.

Consider, for example, an agent that kills cells by inducing apoptosis. That agent favors variant cell lineages that resist the induction of apoptosis. Clonal expansion of the anti-apoptotic lineages follows. Anti-apoptosis may often be an early step in carcinogenesis.

Variant cell lineages arise continuously. However, in the absence of a selective agent to expand clones of predisposed cells, variant cell lineages may have relatively little chance of completing progression. In this regard, selective agents may play a key role in raising cancer incidence. As always, variation and selection must complement each other in the evolutionary process of transformation.

HYPOTHESIS

A recent theory proposes that carcinogens may act as both mutagens and selective agents (Breivik and Gaudernack 1999b; Fishel 2001). In the presence of a mutagen that causes a certain type of DNA damage, selection may favor cells that lose the associated repair pathway. Cells that lack the appropriate repair response may not stop the cell cycle to wait for repair or may not commit apoptosis, whereas repair-competent cells often slow or stop their cycle during repair. Thus, repair-deficient cells could outcompete repair-competent cells, as long as the gain in survival or in the speed of the cell cycle offsets any loss in division efficacy caused by the increased accumulation of mutations.

In support of their theory, Breivik and Gaudernack (1999a) noted the association between the physical location of colorectal tumors and the loss of particular types of DNA repair. Proximal colorectal tumors tend to have microsatellite instability caused by loss of mismatch repair (MMR) genes. The MMR pathway repairs damage caused by methylating carcinogens. Breivik and Gaudernack (1999a) argue that methylating carcinogens often arise from bile acid conjugates that occur mainly in the proximal colorectum.

The argument for proximal tumors can be summarized as follows. Methylating carcinogens concentrate in the proximal colorectum. The MMR pathway repairs the damage caused by methylating agents. Those cells that lose the MMR repair pathway gain an advantage in the selective environment created by methylating agents, because MMR-deficient cells slow down less for repair or commit apoptosis less often than do MMR-competent cells.

By contrast, distal colorectal tumors tend to have chromosomal instability caused by loss of the mechanisms that maintain genomic integrity, such as the nucleotide excision repair (NER) pathway. Breivik and Gaudernack (1999a) argue that the bulky-adduct-forming (BAF) carcinogens may arise primarily from dietary and environmental factors and concentrate primarily in the distal colorectum.

The argument for distal tumors can be summarized as follows. BAF carcinogens concentrate in the proximal colorectum. The NER pathway primarily repairs the damage caused by BAF agents. Those cells that lose the NER repair pathway gain an advantage in the selective environment created by BAF agents, because NER-deficient cells slow down less for repair or commit apoptosis less often than do NER-competent cells.

By this theory, a carcinogen may act in three stages. First, direct mutagenesis creates variant cell lineages. Second, selection favors clonal expansion of variant cells that lose repair function for the type of mutagenic damage caused by the carcinogen. Third, direct mutagenesis of cells that lack associated repair processes may speed the rate at which subsequent transitions occur through the steps of multistage progression.

TEST

By *test,* I mean the ways in which to study the predicted consequences of a carcinogen for age-specific incidence. This section focuses on carcinogens that may act both as direct mutagens and as selective agents. No clear theory has been defined to formulate hypotheses for the relation between the dosage of such carcinogens and the patterns of age-specific incidence.

I can speculate a bit. As mentioned above, a directly mutagenic agent may have three separate effects: initial mutagenesis, secondary selective expansion of mutator clones, and tertiary mutagenesis.

Consider a particular mutagen and an associated DNA repair system that fixes the kind of damage caused by the mutagen. A knockout genotype with reduced or absent repair function should respond differently to the carcinogen when compared to the wild type. In particular, the incidence rate of the knockout should be insensitive to the initial mutagenesis directed at the repair system under study, because that repair system has already been mutated in the germline. The knockout should also be insensitive to clonal expansion, because in the knockout all cells share the loss of repair function and so there should be no selective advantage for relative loss of repair function. The knockout should be affected mainly by the tertiary mutagenesis.

Quantitative predictions could be developed for the relative incidence patterns in wild-type and knockout genotypes, using the methods of the earlier theory chapters. Those predictions could be tested in laboratory animals. Although such tests may not be easily accomplished, it is worthwhile to consider how to connect carcinogenic effects to mechanism, and mechanism to incidence. Ultimate understanding of cancer can only be achieved by understanding how factors influence the rates of progression, and how rates of progression affect incidence.

9.4 Summary

This chapter analyzed classical explanations for chemical carcinogenesis. Those explanations focused on how dosage and duration of chemical exposure may alter incidence. The classical explanations are not as compelling as they originally appeared. The problem arises from the ease with which alternative models can be fit to the data. To avoid the problems of fitting models to the data, I showed how one may frame quantitative hypotheses about chemical carcinogenesis as comparative predictions—the most powerful method for testing causal interpretations of cancer progression.

The next chapter turns to mortality patterns for the leading causes of death. I show that the quantitative tools I have developed to study cancer may help to understand the dynamics of progression for other age-specific diseases and the processes of aging.

10 Aging

This chapter analyzes age-specific incidence for the leading causes of death. I discuss the incidence curves for mortality in light of multistage theories for cancer progression. This broad context leads to a general multicomponent reliability model of age-specific disease.

The first section describes the age-specific patterns of mortality for the twelve leading causes of death in the USA. Heart disease and various other noncancer causes of death share two attributes. From early life until about age 80, the acceleration in mortality increases in an approximately linear way. After age 80, mortality decelerates sharply and linearly for the remainder of life. By contrast, cancer and a couple of other causes of death follow a steep, nearly linear rise in mortality up to 40–50 years, and a steep, nearly linear decline in acceleration later in life. The late-life deceleration of aggregate mortality over all causes of death has been discussed extensively during the past few years (Charlesworth and Partridge 1997; Horiuchi and Wilmoth 1998; Pletcher and Curtsinger 1998; Vaupel et al. 1998; Rose and Mueller 2000; Carey 2003).

The second section presents two multistage hypotheses that fit the observed age-specific patterns of mortality. The increase in acceleration through the first part of life may be explained by a slow increase in the transition rate between stages—perhaps a slow increase in the failure rate for components that protect against disease. With regard to the late-life decline in acceleration, all multistage models produce a force that pushes acceleration down at later ages. That downward force comes from the progression of individuals, as they grow older, through the early stages of disease.

The third section expands the multistage theory of cancer to a broader reliability theory of mortality. For cancer, genetic and morphological observations support the idea that tumor development progresses through a sequence of stages. For other causes of death, little evidence exists with regard to stages of progression. A multicomponent reliability framework seems more reasonable: the reliability (lifespan) of organisms may depend on the rates of failure of various component subsystems that together determine disease progression and survival. Multistage progression corresponds to multiple components arranged in a series. By contrast, functionally redundant components act in parallel; disease arises when all components fail independently.

In the final section, I argue that my extensive development of multistage theory for cancer provides the sort of quantitative framework needed to apply reliability theory to mortality. For cancer, I have shown how multistage theory leads to many useful hypotheses: the theory predicts how age-incidence curves change in response to genetic perturbations (inherited mutations) and environmental perturbations (mutagens and mitogens). Reliability theory will develop into a useful tool for studies of mortality and aging to the extent that one can devise testable hypotheses about how age-incidence curves change in response to measurable perturbations.

10.1 Leading Causes of Death

Figure 10.1 illustrates mortality patterns for non-Hispanic white females in the United States for the years 1999 and 2000. The top row of panels shows the age-specific death rate per 100,000 individuals on a log-log scale. The columns plot all causes of death, death by heart disease, and death by cancer.

The curves for death rate in the top row have different shapes. To study quantitative characteristics of death rates, it is useful to present the data in a different way. The second row of panels shows the same data, but plots the age-specific acceleration of death instead of the age-specific rate of death. The log-log acceleration (LLA) is simply the slope of the rate curve in the top panel at each age. Plots of acceleration emphasize how changes in the rate of mortality vary with age (Horiuchi and Wilmoth 1997, 1998; Frank 2004a).

The bottom row of panels shows one final plotting transformation to aid in visual inspection of mortality patterns. The bottom row takes the plots in the row above, transforms the age axis to a linear scale to spread the ages more evenly, and applies a mild smoothing algorithm that retains the same shape but smooths the jagged curves. I use the transformations in Figure 10.1 to plot mortality patterns for the leading causes of death in Figure 10.2, using the style of plot in the bottom row of Figure 10.1.

Figure 10.2 illustrates the mortality patterns for non-Hispanic white males in the United States for the years 1999 and 2000. Each plot shows

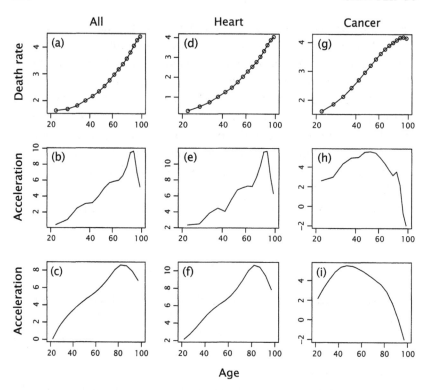

Figure 10.1 Age-specific mortality patterns by cause of death. Data averaged for the years 1999 and 2000 for non-Hispanic white females in the United States from statistics distributed by the National Center for Health Statistics, http://www.cdc.gov/nchs/, Worktable Orig291. The top row of panels shows the age-specific death rate per 100,000 individuals on a log-log scale. The columns plot all causes of death, death by heart disease, and death by cancer. The second row of panels shows the same data, but plots the age-specific acceleration of death instead of the age-specific rate of death. Acceleration is the derivative (slope) of the rate curves in the top row. The bottom row takes the plots in the row above, transforms the age axis to a linear scale to spread the ages more evenly, and applies a mild smoothing algorithm that retains the same shape but smooths the jagged curves. From Frank (2004a).

a different cause of death and the percentage of deaths associated with that cause.

The panels in the left column of Figure 10.2 show causes that account for about one-half of all deaths. Each of those causes shares two attributes of age-specific acceleration. From early life until about age 80, the acceleration in mortality increases in an approximately linear way.

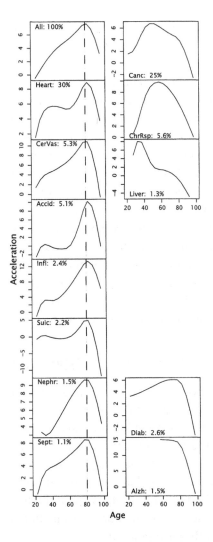

Figure 10.2 Age-specific acceleration of mortality by cause of death. Data averaged for the years 1999 and 2000 for non-Hispanic white males in the United States from statistics distributed by the National Center for Health Statistics http://www.cdc.gov/nchs/, Worktable Orig291. The causes of mortality are based on the International Classification of Diseases, Tenth Revision http://www.cdc.gov/nchs/about/major/dvs/icd10des.htm. The diseases are: *Heart* for diseases of the heart; *CerVas* for cerebrovascular diseases; *Accid* for accidents (unintentional injuries); *Infl* for influenza and pneumonia; *Suic* for intentional self-harm (suicide); *Nephr* for nephritis, nephrotic syndrome and nephrosis; *Sept* for septicemia; *Canc* for malignant neoplasms; *ChrRsp* for chronic lower respiratory diseases; *Liver* for chronic liver diseases and cirrhosis; *Diab* for diabetes mellitus; and *Alzh* for Alzheimer's disease. From Frank (2004a).

After age 80, acceleration declines sharply and linearly for the remainder of life. Some of the causes of death also have a lower peak between 30 and 40 years.

The panels in the upper-right column of Figure 10.2 show causes that account for about one-third of all deaths. These causes follow steep, linear rises in mortality acceleration up to 40–50 years, and then steep, nearly linear declines in acceleration for the remainder of life. The bottom-right column of panels shows two minor causes of mortality that are intermediate between the left and upper-right columns.

What can we conclude from these mortality curves? The patterns by themselves do not reveal the underlying processes. However, the patterns do constrain the possible explanations for changes in age-specific mortality. For example, any plausible explanation must satisfy the constraint of generating an early-life rise in acceleration and a late-life decline in acceleration, with the rise and fall being nearly linear in most cases. A refined explanation would also account for the minor peak in acceleration before age 40 for certain causes.

10.2 Multistage Hypotheses

The mortality curves show a rise in acceleration to a mid- or late-life peak, followed by a steep and nearly linear decline at later ages.

In earlier chapters, I provided an extensive analysis of multistage models. Within the multistage framework, many alternative assumptions can often be fit to the same age-incidence pattern. Thus, fits to the data can only be regarded as a way to generate specific hypotheses. With that caveat in mind, I describe some multistage assumptions that fit the mortality curves and thus provide one line for the development of particular hypotheses (Frank 2004a).

Several alternative models may cause a rise in acceleration through the first part of life. Perhaps the simplest alternative focuses on the transition rates between stages of progression. If transition rates increase slowly with age, then acceleration will rise with age (Figures 6.8, 6.9).

With regard to the late-life decline in acceleration, all multistage models produce a force that pushes acceleration down at later ages. That downward force comes from the progression of individuals, as they grow older, through the early stages of disease (Figures 6.1, 6.2).

If, for example, n stages remain before death, then the predicted slope of the log-log plot (acceleration) is $n - 1$. As individuals age, they tend to progress through the early stages. If there are n stages remaining at birth, then later in life the typical individual will have progressed through some of the early stages, say a of those stages. Then, at that later age, there are $n - a$ stages remaining and the slope of the log-log plot (acceleration) is $n - a - 1$. As time continues, a rises and the acceleration declines (Frank 2004a, 2004b).

10.3 Reliability Models

For cancer, I have been using various stepwise multistage models. Those stepwise models were originally developed for cancer in the 1950s (see Chapter 4) based on the idea of a sequence of changes to cells or tissues, for example, a sequence of somatic mutations in a cell lineage. Later empirical research has supported stepwise progression, based on both genetic and morphological stages in tumorigenesis.

Cancer researchers sometimes argue about what kinds of changes to cells and tissues determine stages in progression, the order of such changes, the number of different pathways of progression for a given type tissue and tumor, and how many rate-limiting changes must be passed for carcinogenesis. But those arguments take place within the multistage framework, which provides the only broad theoretical structure for studies of cancer. The multistage framework developed internally within the history of cancer research, with relatively little outside influence. For those reasons, I have presented the multistage theory with reference only to cancer.

By contrast, studies of heart disease and other causes of mortality face different biological problems and have a different theoretical tradition. On the biological side, most diseases do not have widely accepted stages of progression or widely accepted processes, such as somatic mutation, that drive transitions between stages. Certainly, some multistage progression ideas exist for noncancerous diseases (Peto 1977), and some theories about somatic mutation have been posed (e.g., Andreassi et al. 2000; Vijg and Dolle 2002; Kirkwood 2005; Wallace 2005; Bahar et al. 2006). But those ideas and theories do not form a cohesive framework in current studies of mortality.

Several theories of age-specific mortality have been based on multiple stages or multiple states of progression. Specific models almost always derive from reliability theory—the engineering field that evaluates time to failure for manufactured devices (Gavrilov and Gavrilova 2001).

In engineering, components of a device that protect against failure may be arranged in various pathways. Serial protection means that system failure follows a pathway in which first one component fails, followed by a second component, and so on; the probability of failure of later components in the sequence occurs conditionally on the failure of earlier components in the sequence. Parallel protection describes functional redundancy, in which any single functioning component keeps the system going; failure occurs only after all redundant components fail; and component failures occur independently. Various combinations of serial and parallel pathways may be designed.

Reliability theory calculates time to failure (mortality) based on assumptions about component failure rates and pathways by which components are related. Obviously, the multistage theory I developed earlier forms a branch of reliability theory. However, the reliability theory found in texts focuses on engineering problems, and those problems rarely match the particular biological scenarios for cancer progression. So, although the principles exist in reliability texts, many of the specific results in my theory chapters are new.

Gavrilov and Gavrilova (2001) provided a nice review of reliability theory applied to human mortality. They note that when system failure depends on the simultaneous failure of several components, the acceleration of age-specific mortality declines later in life. I have already discussed the idea several times. If system failure requires failure of n components, then log-log acceleration (LLA) is $n - 1$. As systems age and components fail, say a have failed, then LLA tends to drop toward $n - a - 1$. Details vary, but the idea holds widely. Vaupel (2003) gives a good, intuitive description of how multicomponent reliability may explain the late-life mortality plateau.

In light of reliability theory, we can state more generally an explanation for the late-life decline in the acceleration of mortality (Frank 2004a). Suppose a measurable disease outcome, such as death, occurs only after several different rate-limiting events have occurred. Each event has at least some aspect of its time course that is independent of other events. If so, then the dynamics of onset will not follow the course for a single event model, and will instead be the outcome of a multi-event model. The events do not have to follow one after another or be arranged in any particular pattern. The key is at least partial independence in the time course of progression for each event, and final measured outcome (mortality) only occurring after multiple events have occurred.

Similarly, a condition for a midlife rise in acceleration is a slow increase in the rate at which individual components fail (Frank 2004a).

10.4 Conclusions

I have included a discussion of mortality in a book otherwise devoted to cancer for two reasons. First, from the vantage point of the general reliability problem, one can more easily see what is necessary to explain patterns of cancer incidence. Second, the extensive development of multistage theory I presented in earlier chapters provides just the sort of quantitative background needed to use reliability theory fruitfully in the general study of mortality.

One might now ask: If reliability theory applies to everything, then does it have any explanatory power? This question seems reasonable, but I think it is the wrong question. The reliability framework provides tools to help us formulate testable hypotheses. That framework by itself is not a hypothesis.

For cancer, I have shown how multistage theory leads to many useful hypotheses. For example, I have used the theory to predict how age-incidence curves change in response to genetic perturbations (inherited mutations) and environmental perturbations (mutagens and mitogens). Reliability theory will develop into a useful tool for studies of mortality and aging to the extent that one can develop useful hypotheses about how age-incidence curves change in response to measurable perturbations.

10.5 Summary

This chapter finishes my three empirical analyses of disease dynamics in light of multistage progression models. The three empirical analyses covered genetics, chemical carcinogenesis, and aging. The next section of the book turns to evolutionary processes: What factors shape the population frequencies of predisposing genetic variants? How does tissue architecture affect the somatic evolution of cancer?

PART III

EVOLUTION

11 Inheritance

Cancer progresses by the accumulation of heritable changes in cell lineages. In the simplest case, all of the changes happen to the DNA of a single somatic cell lineage. Starting with the initial cell, the carcinogenic process develops through the sequential addition of genetic changes that eventually gives rise to the tumor.

Many cancer biologists rightly object to this oversimplified view. The heritable changes may often be epigenetic—genomic changes other than DNA sequence—or physiological changes that persist (inherit) for many cell generations. Changes may happen to multiple lineages, with carcinogenesis influenced by positive feedback between altered lineages. But even this richer view still comes down to heritable changes in cell lineages—almost necessarily so, because cells are the basic units, and persistent change means heritable change. Disease arises at the level of tissues, but the causes derive from changes to cells.

The first heritable carcinogenic changes may trace back to a somatic cell that descended from the zygote, in which case the changes derive purely from the somatic history of that organism. Or the origin of a particular inherited variant may trace back to a germline cell in one of the individual's ancestors, in which case the inherited variant may be shared by other descendants.

All of these descriptions turn on heritable change in lineages, that is, on evolutionary change. Cancer has long been understood in terms of somatic evolution within an individual's cellular population. More recently, the role of inherited germline variants has been studied in terms of the evolutionary genetics of populations of individuals.

We can think about any particular variant, somatic or germline, in two ways. First, the variant influences disease through its effect on progression—the role of development that traces cause from genes to phenotypes. Second, the phenotype influences whether, over time, the variant lineage expands or goes extinct—the role of natural selection in shaping the distribution of variants.

The following chapters focus on variants that originate in somatic cells: in a particular cell, variants trace their origin back to an ancestral cell that descended from the most recent zygote. Somatic variants drive progression within an individual.

This chapter focuses on germline variants that may occur in different individuals in the population: in a particular cell, germline variants trace their origin back to an ancestral cell that preceded the most recent zygote. Germline variants determine inherited predisposition to cancer.

The first section describes how inherited variants affect progression and incidence—the causal pathway from genes to phenotypes. A classical Mendelian mutation is a single variant that strongly shifts age-onset curves to earlier ages. Such mutations demonstrate the central role of inherited variation in progression and the multistage nature of carcinogenesis. Other inherited variants may only weakly shift age-onset curves; however, the combination of many such variants predisposes individuals to early-onset disease.

The second section turns around the causal pathway: the phenotype of a variant—progression and incidence—influences the rate at which that variant increases or decreases within the population. The limited data appear to match expectations: variants that cause a strong shift of incidence to earlier ages occur at low frequency; variants that cause a milder age shift occur at higher frequencies; and variants that only sometimes lead to disease occur most frequently.

The final section addresses a central question of biomedical genetics: Does inherited disease arise mostly from few variants that occur at relatively high frequency in populations or from many variants that each occur at relatively low frequency? The current data clarify the question but do not give a clear answer. Inheritance of cancer provides the best opportunity for progress on this key question.

11.1 Genetic Variants Affect Progression and Incidence

The first studies measured differences in progression and age of onset between variants at a single locus. Those first studies aggregated all variants into two classes, wild type and mutant, and compared incidences between those classes. Current studies measure differences at a finer molecular scale, distinguishing between variants at a particular nucleotide or amino acid site, or between variants that differ by single insertions or deletions. Ultimately, one would like to know how variants at multiple sites combine to affect incidence. So far, most studies have

been limited to indirect analysis of multiple sites by associations between familial relationships and incidence, the classical nonmolecular
approach to quantitative inheritance.

Variants at a Single Locus

This section compares progression and incidence between individuals who carry, at a single locus, either the wild-type allele or a loss of
function mutation. In most cases, one compares homozygotes for the
wild type and heterozygotes that carry one wild-type and one loss of
function mutation. In practice, "wild type" means the class of all variant
alleles that do not have a large effect on incidence, and "loss of function"
means the class of all variant alleles that cause a large increase in the
rate of progression.

The comparison between individuals carrying wild-type and loss of
function genotypes played a key role in the history of multistage theories of carcinogenesis. The shift of the incidence curve to earlier ages
in the loss of function genotypes provided the first direct evidence that
mutations in cell lineages affect progression. The observed magnitude
of the shift in incidence curves matched the expected shift under multistage theory. In that theory, progression follows the accumulation of
multiple genetic changes, and the inherited mutation provides the first
of two or more steps in carcinogenesis.

In earlier chapters, I described studies that compared age incidence
between genotypes that differed at a single locus, comparing the wild-
type with loss of function mutations. In this section, I copy the figures
from two earlier examples. The following sections provide new examples.

Figure 11.1 compares incidence rates between inherited and sporadic
cases of retinoblastoma. In the inherited cases, individuals carry one
mutated allele at the retinoblastoma locus. Within the multistage framework, inheriting a key mutation means being born one stage advanced in
progression. The theory predicts that an advance by one stage reduces
the slope of the incidence curve by one. The difference in the log-log acceleration (LLA) of the two incidence curves measures the difference in
the slopes of the incidence curves. Figure 11.1c shows that the observed
difference in slopes is close to one, matching the theory's prediction.

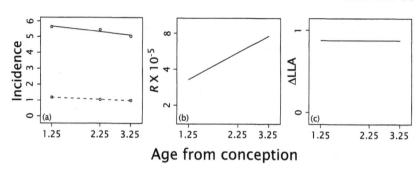

Figure 11.1   Age-specific incidence of bilateral and unilateral retinoblastoma. Bilateral cases are mostly inherited, and unilateral cases are mostly sporadic. (a) Bilateral (solid line) and unilateral (dashed line) incidence of retinoblastoma per 10^6 population, shown on a $\log_{10}$ scale. (b) Ratio, R, of unilateral to bilateral incidence at each age multiplied by 10^{-5}, using the fitted lines in the previous panel. (c) Difference in log-log acceleration between unilateral and bilateral cases, which is the log-log slope of R versus age in Eq. (8.2). Ages measured in years. I presented this figure earlier as Figure 8.3; see my earlier presentation for more details.

Figure 11.2 compares incidence rates between inherited and sporadic cases of colon cancer. In the inherited cases, individuals carry one mutated allele at the *APC* locus. Again, the multistage framework predicts that an inherited mutation in a key rate-limiting process advances progression by one stage and therefore reduces the log-log acceleration of incidence by one. Figure 11.2c shows a difference in LLA of about 1.5, a reasonable match to the theory's prediction given the sample sizes and complexities of progression.

Common Variants at a Single Site

The previous section described studies that aggregated variants into wild-type and mutant classes. This section presents two cases in which mutations at specific sites define the variants.

BRCA Mutations and Breast Cancer

Struewing et al. (1997) screened Ashkenazi Jewish females for two specific mutations in *BRCA1* and one specific mutation in *BRCA2*. They obtained age of breast cancer onset among the 89 carriers and 3653 noncarriers. They used a statistical procedure that accounted for relatedness between certain sample members to obtain estimates for the risk of breast cancer, measured as the expected fraction of women at each

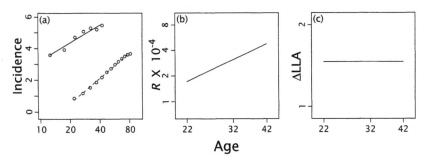

Figure 11.2 Age-specific incidence of inherited familial adenomatous polypo-sis (FAP) and sporadic colon cancer. (a) Inherited colon cancer (FAP) caused by mutation of the *APC* gene (top curve) and sporadic cases (bottom curve) per 10^6 population, shown on a $\log_{10}$ scale. (b) Ratio, *R*, of sporadic colon cancer incidence to inherited FAP incidence at each age multiplied by 10^{-4}, using the data in the previous panel. (c) The difference in the log-log acceleration between sporadic and inherited cases, which is the log-log slope of *R*. I presented this figure earlier as Figure 8.5; see my earlier presentation for more details.

five-year age interval who would be expected to develop cancer by that age.

In Figure 11.3a, the circles plot their estimates, shown as the fraction who would be expected not to have developed a breast tumor by each age. The solid curve provides a smoothed fit to the carrier class; the dashed curve provides a smoothed fit to the noncarrier class.

In the data from Struewing et al. (1997), the estimated fraction tumor-less sometimes increases from one age to a later age. Such increases are, of course, not possible in the actual fraction tumorless curves. The increases arise because of the estimation procedure. I mention this be-cause the rise and fall in the estimates (shown as circles) at later ages causes the curves to be particularly sensitive to the smoothing param-eters. For these reasons, and the moderately small sample of carriers, these data only illustrate various ways in which to analyze such prob-lems.

With current technological trends, we will eventually have vastly more data of this kind. At present, I focus mainly on exploratory analysis to highlight some interesting hypotheses, which will require further stud-ies to test.

Hypothesis 1: All carriers do not have highly elevated risk.—The second row of panels in Figure 11.3 plots the standard log-log incidence curves

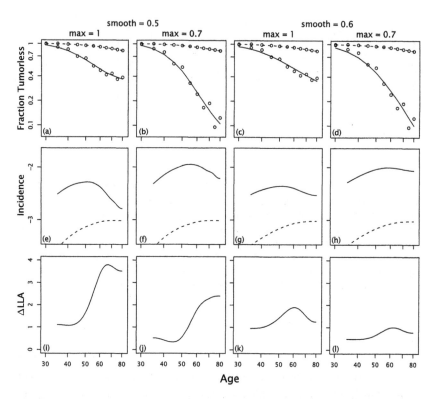

Figure 11.3   Breast cancer rates for females who carry a mutation in *BRCA1* or *BRCA2*, shown as solid lines, versus those females who do not have a mutation, shown as dashed lines. The circles in (a) and (c) mark the estimated fraction of females in each class that have not yet developed tumors, taken from Figure 1B of Struewing et al. (1997). In (b) and (d), I transformed the fraction tumorless, f, as $S = (max - f)/max$, where *max* is the fraction of the carriers who have fully elevated risk. Panels (a) and (b) used the smooth.spline function of the R computing language (R Development Core Team 2004) to fit a smooth curve to the observed points, with smoothing parameter set to 0.5; (c) and (d) force a stiffer, less curved fit with a smoothing parameter of 0.6. The second row shows incidence on a $\log_{10}$ scale, obtained from $-d\ln(S)/dt$, where S is the fraction tumorless in the curves of the top row. The bottom row shows ΔLLA, the difference in the log-log slopes of incidence in the second row of plots.

for carriers and noncarriers. In all four panels, the noncarriers (dashed curve) show the commonly observed pattern for sporadic breast cancer: a diminishing slope of incidence with age, but little or no actual decrease in the incidence rate before age 80. By contrast, the incidence declines after midlife for the carriers (solid curves) in all of the panels except panel (h). I work through the steps that lead to panel (h). As I mentioned, I do not regard these manipulations as tests of any hypothesis, but rather as ways to generate new hypotheses.

Panel (e) shows the direct estimate of carrier incidence using the original values of Struewing et al. (1997) and the standard smoothing parameter of 0.5 for fitting the curves in panel (a). In (e), carrier incidence declines strongly and steadily after about age 55. In (f), I considered the possibility that only a fraction of carriers have highly elevated risk. The division of carriers into very high risk and moderate risk categories may arise from genetic predisposition caused by other loci. I discuss evidence for this idea in following sections; here I just look at the consequences.

The estimated fraction of carriers who develop cancer by age 80 is about 0.66. What if nearly all carriers with highly elevated risk develop cancer? Suppose, for example, that only a fraction $max = 0.7$ of carriers have elevated risk, and nearly all of them develop cancer. Then the fraction tumorless among the class with highly elevated risk is $S = (max - f)/max$, where f is the fraction tumorless among all carriers. Panels (b) and (d) show the fraction tumorless among carriers with highly elevated risk, using $max = 0.7$. Panel (f), derived from (b), has a carrier incidence curve that drops later in life, but less strongly than in (e).

Panel (h), derived from (d), has what I consider to be the right shape for the carrier incidence curve. The difference between (h) and (f) comes only from the smoothing parameter used to fit the curves in the top row. Whenever a key match to expectations arises only from a moderate change in the smoothing parameter, one clearly does not have enough data to draw any conclusions. Normally, after seeing such a pattern, I would suggest not presenting such an analysis. I present it here to warn about the importance of sample size and sensitivity to smoothing procedures, and because I think the alternative biological interpretations are sufficiently interesting to stimulate further work.

In summary, I suggest that the estimated incidence curve in (h), based on the stiffer smoothing method, comes closer to the actual incidence pattern. More importantly, I propose that, among carriers, only a fraction have highly elevated risk. I will discuss below two ways in which background genotype may elevate risk in some *BRCA* mutant carriers.

Hypothesis 2: BRCA mutations abrogate a rate-limiting step.—An inherited mutation may increase incidence in at least two different ways.

First, an inherited mutation may raise the rate of somatic mutations, including epigenetic and chromosomal changes. In this case, the inherited mutation may not abrogate a rate-limiting step, but instead increase the transition rates between the normal rate-limiting steps that characterize carcinogenesis in the absence of the mutation. If so, then the theory predicts a rise with age in the difference between the log-log slopes of incidence (ΔLLA) for sporadic versus inherited cases. (See Eq. (7.6) and Figures 7.5 and 7.6.)

Second, an inherited mutation may directly or indirectly abrogate a single rate-limiting step. In this case, the theory in Eq. (7.5) predicts that ΔLLA ≈ 1 and does not change much with age.

The bottom row of Figure 11.3 shows a range of patterns for ΔLLA. In panel (i), the value rises strongly with age; in panel (l), the value remains mostly flat and near one. The two middle panels follow intermediate trends. We do not know enough yet to assign significantly higher likelihood to one pattern over the others because of: the limited sample size for inherited cases; the fluctuations in the fraction tumorless caused by the estimation procedure in the original paper; and the uncertainty with regard to the fraction of carriers who have elevated risk.

I favor the right column of panels in Figure 11.3, because the incidence pattern for carriers has the common shape for breast cancer, in which incidence plateaus later in life but does not decline significantly before age 80. The right column matches the prediction for a *BRCA* mutation to knock out one rate-limiting step. To test that hypothesis, we need more data on incidence in carriers and on the fraction of carriers who have highly elevated risk.

MDM2 VARIANT AND THE P53 PATHWAY

p53 is the most commonly mutated gene in tumors. In some tumors, mutations arise in those genes that regulate p53 rather than in *p53* itself.

To search for new inherited variants that affect the p53 system and cancer, Bond et al. (2004) focused on MDM2, a direct negative regulator of p53. They found a single nucleotide polymorphism in the *MDM2* promoter that enhanced MDM2 expression and attenuated the p53 pathway. In particular, the variant had a T → G change at the 309th nucleotide of the first intron (SNP309). This SNP occurred at high frequency in a sample of 50 healthy individuals: heterozygote T/G at 40% and homozygote G/G at 12%.

A variant affects cancer to the extent that it shifts the age-onset curve to earlier ages. To measure the variant's effect, Bond et al. (2004) studied a group that suffered soft tissue sarcoma (STS) and had no known *p53* or other predisposing inherited mutations.

The data collected by Bond et al. (2004) show that the variant allele shifts age of onset to earlier ages, supporting the hypothesis that the variant's increased expression of MDM2 enhances tumor progression. However, Bond et al.'s (2004) particular quantitative analyses misuse the data and the theory of multistage progression. I demonstrate proper analysis, because this study provides just the sort of combined genetic, functional, and population level insight that will be required to move the field ahead.

Figure 11.4a,b presents copies of Figure 7C,E from Bond et al. (2004). Panel (a) compares age of onset for all soft tissue sarcomas between the wild type (T/T) and the homozygote variant (G/G). The wild type progresses at a median age of 59 compared with a median of 38 for the homozygote variant, showing the earlier onset for the variant.

In the sample collected by Bond et al. (2004), liposarcomas form the largest subset of soft tissue sarcomas. Figure 11.4b shows how Bond et al. (2004) fit curves to the onset data for liposarcoma in order to estimate the number of rate-limiting steps in progression for each genotype. They assumed that the y axis measured incidence, and fit $I(t) = kt^{n-1}$ (they used r instead of n for the number of rate-limiting steps). From their fitting procedure, they estimated n as 4.8 for the wild type (T/T, solid curve), 3.5 for the heterozygote (T/G, dashed curve), and 2.5 for the homozygote variant (G/G, dot-dash curve). These estimates differ by about one, so the authors concluded that the variant abrogates one rate-limiting step in progression. I do not know whether the biological conclusion is correct, but the analysis of the data is inappropriate.

The y axis of Figure 11.4b measures the percentage of individuals of a particular genotype who have suffered cancer by a particular age. That measure differs from incidence. I have shown previously that such data can be transformed into incidence. Let y be the percentage of individuals with cancer by age t, as on the y axis of Figure 11.4b. Then the fraction tumorless is $S = 1 - y/100$, where the 100 arises because y is given as a percentage. Incidence is $I(t) = -d\ln(S)/dt$.

To study the curves in relation to the number of rate-limiting steps, n, we can use the form applied by Knudson (1971), $\ln(S) = -k_1 t^n$, where k is a constant, or, differentiating $\ln(S)$ with respect to t, we can use incidence, $I(t) = k_2 t^{n-1}$. I discussed in earlier chapters the theory behind these equations.

If I were to analyze the data in Figure 11.4b, I would highlight two issues before starting. First, there are only four individuals in the variant homozygote (G/G) sample. One will not get a reliable estimate of a rate (incidence) from four observations. Second, the median age of onset is nearly identical for the wild type (T/T) and the heterozygote (T/G). Median age of onset often provides a good measure for the rate of progression as, for example, in the classical Druckrey analysis of chemical carcinogens (see Section 2.5). With nearly identical medians for those two genotypes, I would not be inclined to put much weight on any estimated differences in the slopes of the incidence curves, unless I had reason to believe that one genotype had both more rate-limiting steps and a faster transition rate between steps than the other genotype. In this study, those assumptions would over-interpret the data.

Given these issues with regard to the data analysis of Figure 11.4b, I would be content to note that the direction of shift in the homozygote variant (G/G) is consistent with enhanced progression.

I have emphasized data interpretation because the work of Bond et al. (2004) is just the sort of study that will become increasingly common and important as genomic technology improves. I agree with the authors that the analysis of inherited variants comes down to understanding how those variants affect age of onset. Further, the quantitative aspects of rates could, in principle, provide insight into the mechanisms by which variants influence the complex process of progression. With the inevitably larger samples that will soon be available, it should be possible to accomplish such analyses with much greater ease and power.

Interaction between Variants at Different Sites

Variants at different nucleotide sites may interact to influence progression. Studies to date have generally not had sufficient resolution and sample sizes to demonstrate the joint effects of different variants on age-incidence patterns in human populations. The work of Bond et al.

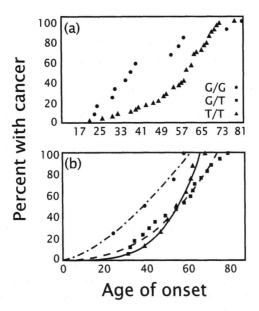

Figure 11.4 Onset of soft tissue sarcoma for individuals classified by genotype at a single nucleotide polymorphism in the promoter region of *MDM2*. At the polymorphic site, individuals are wild type (T/T), heterozygote for the variant allele (T/G), or homozygote for the variant (G/G). The y axis shows the percentage of individuals of a particular genotype who have suffered cancer by a particular age. (a) The homozygote variant has earlier age of onset than the wild type. (b) Pattern for those soft tissue sarcomas classified as liposarcoma, the most common form of soft tissue sarcoma in the sample. Redrawn from Figure 7C,E of Bond et al. (2004).

(2004) discussed in the previous section provides a glimpse of the sort of study that will become common in the future.

In the previous section, I described how MDM2 acts as a negative regulator of p53. Bond et al. (2004) showed that a nucleotide variant in the promoter of *MDM2* enhances expression of the MDM2 protein and thus negatively influences the p53 regulatory system. In individuals with a normal *p53* locus, the *MDM2* promoter variant enhances progression of soft tissue sarcomas, the same type of cancer often found in individuals who inherit *p53* defects.

Bond et al. (2004) extended their study to samples that included individuals who carry both the *MDM2* promoter variant and a mutation in *p53*. Those double mutant individuals suffered faster progression than individuals who inherited only one of the two mutations. If we use +

and – superscripts to label the wild type and variant, then the ordering of the median age of onset was $MDM2^-/p53^- < MDM2^+/p53^- < MDM2^-/p53^+ < MDM2^+/p53^+$, with values for the medians of $2 < 14 < 38 < 57$.

The *MDM2* variant alone shifts the median from 57 in the wild type to 38; the *p53* variant alone shifts the median from 57 in the wild type to 14. In this case, either variant by itself causes significantly enhanced progression. In other cases, a variant by itself may have little effect in the absence of a synergistic variant at another site.

Comparison between Rare Variants at Single Sites

Technical advances in DNA sequencing efficiency provide an opportunity to study individual nucleotide variants. Ideally, one would like to associate nucleotide variants to their consequences for cancer, measured by the age of cancer onset. However, each particular variant often occurs only rarely in natural populations, so it may be difficult to compare the age of onset between those individuals with and without the variant. In addition, many amino acid substitutions may have a weak effect on biochemical function, whereas a few substitutions may have a strong effect. Some a priori way of weighting the expected effects of particular substitutions would greatly enhance the association between DNA sequence variants and their consequences for cancer onset.

The association between the nucleotide sequence of DNA mismatch repair genes and colorectal cancer has been the focus of many recent studies. In those studies, each observed human subject provides an age of cancer onset and information about variant nucleotide sites or amino acid substitutions in the mismatch repair genes. The two problems mentioned above arise when analyzing the data from those studies: each particular variant occurs rarely, and some method must be used to weight the expected consequences of a substitution.

To solve these problems, various computational methods predict the expected functional consequences of amino acid substitutions. One method examines the evolutionary history of a gene, and weights more heavily those substitutions that occur rarely across different species (Ng and Henikoff 2003). The idea is that relatively rare changes must often be more constrained by functional consequences of substitutions, whereas relatively common changes must often have relatively few deleterious consequences. Another method, polymorphism phenotyping (PolyPhen), combines evolutionary conservation with various measures of biochemical structure and function (Ramensky et al. 2002).

I obtained two unpublished collections of PolyPhen scores for mismatch repair gene variants and the associated ages of colorectal cancer onset. Figure 11.5 presents a preliminary analysis of those data. I particularly wish to emphasize the importance of using the full age of onset data. Many analyses simply classify age of onset as early or late, throwing out the most valuable quantitative aspect of outcome. I have emphasized throughout this book that age of onset provides the summary measure of outcome when studying how various causal factors influence cancer progression.

Figure 11.5a shows the association between single amino acid substitutions and age of onset. These data came from a survey of the literature, in which each publication usually reported a single amino acid variant believed to influence mismatch repair function and age of cancer onset. These confirmed variants form a generally accepted set of DNA repair variants with functional consequences on which we could test the efficacy of the PolyPhen scoring method.

The raw data for Figure 11.5a scatter widely, because so many factors influence the age of cancer onset for each individual case. I used a sliding window analysis to illustrate the strong trend in the data (see figure legend). The result shows a clear tendency for increased PolyPhen score to predict the association between a substitution and the rate of cancer progression measured by age of onset.

The confirmed variants in Figure 11.5a generally had some independent evidence that suggested functional consequence for DNA repair and cancer. If PolyPhen does indeed provide a computational method for predicting consequence, then the method should also work on nucleotide sequences obtained without any a priori information about the functional consequence of variant sites.

Figure 11.5b shows unpublished data collected from individuals for whom early-onset colorectal cancer runs in their family. For each individual, I received the age of colorectal cancer onset and the average PolyPhen score over all 34 variant amino acid sites in the data set. I excluded 26 individuals who did not have any variants and so did not have a predictive PolyPhen score. The remaining 62 individuals each had

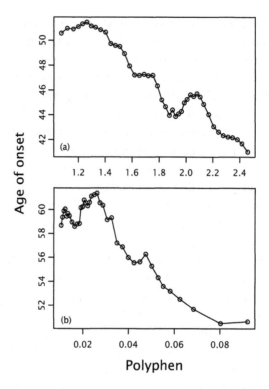

Age of onset

Polyphen

Figure 11.5   Association between cancer onset and the predicted functional consequences of amino acid substitution in DNA repair genes measured by the PolyPhen score. (a) A data set of 78 individuals culled from the literature, in which each paper reported the age of onset and the associated amino acid substitution. The PolyPhen score was calculated for the single amino acid replacement. Each observation provided a PolyPhen score and an age of onset. I first sorted the observations by PolyPhen score. I then calculated a sliding window of average values with a window size of 35. Each point in the figure shows the average value of age and PolyPhen score in the window. (b) Individuals who come from families with a tendency for early-onset colorectal cancer. For each individual, DNA sequences were obtained from parts of the mismatch repair genes *Mlh1, Mlh2,* and *Mlh6.* I used a window size of 25 for this analysis of the 62 individuals with nonzero PolyPhen scores. I obtained these data and all calculations of PolyPhen scores from the laboratory of Steven M. Lipkin at the University of California, Irvine.

one or a few variant sites. The sliding window analysis in Figure 11.5b demonstrates the predictive power of the PolyPhen scoring for age of onset. In this case, the variants were collected blindly with regard to prior knowledge about the functional consequences of particular amino acid substitutions.

Many factors influence age of onset, so the PolyPhen scoring on single variants will provide only a small amount of information about predicted risk and age of onset. The value of the analysis may come from hypotheses about which amino acid sites and which kinds of biochemical function affect DNA repair efficacy, and how those changes in efficacy influence cancer progression. Such hypotheses could be tested in laboratory animals, in which one could construct genotypes with particular amino acid substitutions.

Combined Effect of Variants at Multiple Sites

Cancer often aggregates in families, suggesting a strong inherited component that predisposes individuals to disease. In two well-studied cancers, breast and colon, only about 10–20% of the inherited component can be explained by known variants (Anglian Breast Cancer Study Group 2000; de la Chapelle 2004). Those known variants include *BRCA1* and *BRCA2* for breast cancer and *APC* and the mismatch repair genes for colon cancer. Each of those variants causes a large change in the incidence curve. The large effect of such variants makes them relatively easy to study: compare the incidence curves between genotypes with and without the variant. A small sample provides sufficient power to observe the large effect.

Many other variants, each with small effect on incidence, may also occur. However, finding such variants is difficult. One must first identify a candidate variant, and then compare incidence between genotypes with and without the variant in large samples. Such studies remain beyond what can easily be accomplished, even with advancing technology.

STATISTICAL STUDIES OF INHERITANCE

In the absence of direct knowledge about many genes that predispose to cancer, statistical studies have analyzed how environmental and genetic variation contribute to differences in cancer risk. For example, reflecting environmental effects, immigrants take on the risk of colon cancer that is specific for their new home (Haenszel and Kurihara 1968). The risk of developing colon cancer for an individual in a specific geographical region is strongly associated with levels of meat consumption (Armstrong and Doll 1975), so changes in diet might explain the altered risk of immigrants. Smoking (Doll 1998; Vineis et al. 2004) and long-term exposure to certain carcinogens (Vineis and Pirastu 1997) also cause significant environmental risk.

To determine the genetic component of risk, statistical studies compare the frequencies of cancer occurrence between monozygotic twins, dizygotic twins, other family members, and unrelated individuals (Lichtenstein et al. 2000). In principle, such studies could separate the contributions of shared genes, shared environment in the family, and differences in environment between unrelated individuals. However, the statistical power of such studies tends to be low, with wide confidence intervals for the relative roles of genes and environment. This problem is particularly severe for the rarer cancers because of low sample sizes in such studies.

A large study from the Swedish Family-Cancer database provided narrower confidence intervals for the proportions of cancer variance that are explained by genes and environment (Czene et al. 2002). The estimates for genetic contribution ranged from 1% to 53%, depending on the type of cancer. These values may be lower limits, because certain types of genetic variation could not be separated from the effects of a shared environment. Confounding components include similar genotypes between parents, which would be classed as a shared environmental effect rather than a genetic effect. In this study, Mendelian loci explain only part of the total genetic contribution to cancer risk, indicating a significant role for polygenic variation.

An interesting analysis of the Anglian Breast Cancer Study Group study took a different approach to genetic predisposition (Pharoah et al. 2002). The authors first removed the two known Mendelian loci associated with breast cancer—*BRCA1* and *BRCA2*—from the analysis, and then fitted the remaining risk distribution to a polygenic model in which the small risks per variant allele are multiplied across loci. According to the fitted model, the 20% of the population that has the highest level of genetic predisposition has a 40-fold greater risk than the 20% of the population with the lowest level of predisposition. The model also predicted that more than 50% of breast cancers occur in the 12% of the population with the greatest predisposition. The known Mendelian loci account for only a small proportion of the total genetic risk, with the remainder being explained by polygenic variation.

It is difficult to tell how reliable those conclusions are about polygenic inheritance. Other models could be fit to the same data, with different contributions of Mendelian loci, polygenic loci, and environment. I favor the strong emphasis on polygenic inheritance, because most complex quantitative traits in nature show extensive polygenic variation (Barton and Keightley 2002; Houle 1992; Mousseau and Roff 1987). However, statistical models are hard to test directly, because it is difficult to obtain evidence that strongly supports one model and rules out other plausible models. One is often left with conclusions that are based as much on prior belief as on data.

DIRECT STUDIES OF VARIANTS AT MULTIPLE SITES

Ideally, one would like to know how particular genetic variants affect the biochemistry of cells, and how those biochemical effects influence progression to cancer. Although we are still a long way from this ideal, recent studies of DNA repair genes provide hints about what could be learned (Mohrenweiser et al. 2003).

Individuals vary in the ability of their cells to repair DNA damage (Berwick and Vineis 2000). A relatively low repair efficiency is associated with a higher risk of cancer. Presumably, the association arises because higher rates of unrepaired somatic mutations and chromosomal aberrations contribute to faster progression to cancer. Repair genes also play a role in sensing genetic damage and initiating apoptosis.

Most studies of repair capacity measure the effects of mutagens on DNA damage in lymphocytes. For example, a mutagen can be applied to cultures of lymphocytes; after a period of time, damage can be measured by the numbers of unrepaired single-strand or double-strand breaks, or by incorporation of a radioisotope. To study the role of DNA repair in cancer, measurements compare individuals with and without cancer. Berwick and Vineis (2000) summarized 64 different studies that used a variety of methods to quantify repair. In those studies, a relatively low repair capacity was consistently associated with an approximately 2–10-fold increase in cancer risk.

Roughly speaking, repair efficiency has an inheritance pattern that is typical of a quantitative trait. A few rare Mendelian disorders cause severe deficiencies in repair capacity. Apart from those rare cases, repair capacity shows a continuous pattern of variation and has a significant

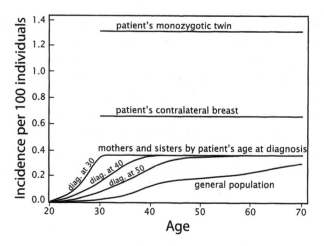

Figure 11.6   Schematic summary of breast cancer incidence in individuals with varying levels of relatedness to an index case. Redrawn from Peto and Mack (2000).

heritable component (Grossman et al. 1999; Cloos et al. 1999; Roberts et al. 1999). Measures of variability and heritability are statistical descriptions of the genetics of repair. Recent studies have made the first steps toward understanding the mechanistic relations between genetic variants and altered phenotypes.

Many genes in the five key repair pathways for different types of DNA damage are known (Bernstein et al. 2002; Thompson and Schild 2002; Mohrenweiser et al. 2003), so genetic variants can be identified by sequencing the loci involved. Specific variants can also be constructed, and their physiological consequences tested in cell-based assay systems. Mohrenweiser et al. (2003) list 22 genes in the core pathway of the MMR system. This system primarily corrects mismatches and short insertion or deletion loops that arise during replication or recombination (Hsieh 2001). The MMR system increases the accuracy of replication by a factor of 100–1,000.

Eighty-five different variants have been found in seventeen different MMR genes that were screened in at least fifty unrelated individuals (Mohrenweiser et al. 2003). Of those variants, 38% occurred at a frequency of 2% or more; 21% occurred at a frequency of 5% or more; and 12% occurred at a frequency of 20% or more. The other DNA repair pathways provided similar results, as summarized by Mohrenweiser

et al. (2003). In 74 repair genes from various pathways, the average frequency of the wild-type allele is approximately 80%, with the remaining 20% comprised of different allelic variants. Among the 148 alleles per person at the 74 repair loci, the average number of allelic variants is expected to be approximately 30. Presumably, each individual carries a very rare or unique genotype.

In summary, small variations in DNA repair are highly heritable, DNA repair efficiency is correlated with cancer risk, and there are widespread amino acid polymorphisms in the known repair genes. The next step will be to link those polymorphisms to variations in the biochemistry of repair, providing a mechanistic understanding of how genetic variation influences an important aspect of cancer predisposition (de Boer 2002).

AGE-SPECIFIC INCIDENCE

The polymorphisms that occur in DNA repair genes hint at variations in cellular physiology that may be very common. The connection between DNA repair efficiency and cancer seems plausible, because somatic mutations and chromosomal aberrations probably have a key role in cancer progression. However, at present, we cannot make a simple mechanistic connection between repair efficacy and the rate of progression to cancer.

Currently, the most interesting studies of multisite variants and age-specific incidence link aggregation of cases in families to age of onset. Presumably, familial cases that rule out known major single-site variants arise from multisite variants shared by relatives.

Peto and Mack (2000) noted that women who are at high risk of developing breast cancer show an approximately constant incidence of cancer per year after a certain age, whereas in most individuals incidence rises significantly with age (Figure 11.6). This pattern appears in three different classes of susceptible individuals after the age at which a particular patient develops cancer. I refer to the individual who first has cancer as the *patient* or the *index case*, and the age of this first diagnosis as the *index age*.

In the first class, an index case with monolateral breast cancer has an annual risk of developing cancer in the other (contralateral) breast of approximately 0.7% per year after the index age. A different study found a similar result, with risk in the contralateral breast of about 0.5% per year after the initial cancer (Figure 11.7).

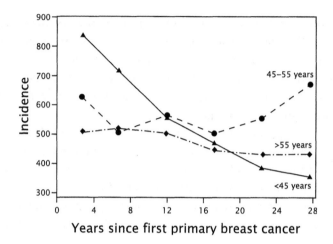

Figure 11.7   Incidence of cancer in the contralateral breast after the first primary breast cancer, excluding cases in which the contralateral cancer was diagnosed within three months of the first cancer. Incidence per year shown on a linear scale per 100,000 population. The earliest cases (solid line) probably carry an excess frequency of *BRCA1* or *BRCA2* mutations (Peto et al. 1999). The decline in incidence for those cases may arise because the subset of individuals who carry *BRCA1* or *BRCA2* mutations may more rapidly develop contralateral tumors. Redrawn from Hartman et al. (2005).

In the second class, a monozygotic twin of an index case has an approximate risk of 1.3% per year after the index age, which is again approximately 0.7% per breast per year.

In the third class, mothers and sisters of an index case have a risk of approximately 0.3–0.4% per year after they have passed the index age.

Single locus mutations of large effect, such as *BRCA1* or *BRCA2*, explain less than one-fifth of familial aggregation (Anglian Breast Cancer Study Group 2000). Thus, the patterns of high and nearly constant incidence most likely arise from familial inheritance of variants at multiple sites—polygenic inheritance.

The tendency for risk after the index age to remain nearly constant for the remainder of life raises an interesting puzzle: what causes that early plateau of incidence in highly susceptible individuals?

HYPOTHESIS FOR EARLY PLATEAU OF INCIDENCE

Peto and Mack (2000) concluded: "A ... model that may account for these peculiar temporal patterns is that many, and perhaps most, breast cancers arise in a susceptible minority whose incidence, at least on average, has increased to a high constant level at a predetermined age that varies between families."

But why should predisposed individuals have constant annual risks after a certain age? Individuals who are not predisposed to breast cancer show an increasing risk with age, and the same is true for the other most common types of epithelial cancer when risk is measured in the absence of information about genetic predisposition.

Frank (2004d) proposed the following explanation for Peto and Mack's (2000) observations. Suppose, at birth, that each of L different cell lineages in the breast has n rate-limiting steps remaining before cancer. I have discussed previously that, as individuals age, their cell lineages may progress independently. Over time, the various lineages form a distribution of stages: some still have n stages remaining before cancer, others have progressed part way and have, for example, $n - a$ stages remaining.

If some cell lineages in an individual have passed through all but the final stage in cancer progression, with only one stage remaining, then that individual's annual risk is constant—the risk is just the constant probability of passing to the final stage. Families that have an increased predisposition may progress through the first $n - 1$ stages quickly; subsequently, their annual risk is the constant probability of passing the final stage. Families with low genetic risk move through the early stages slowly: in middle or late life, members of those families typically have more than one stage to pass and so continue to have an increasing rate of risk with advancing age.

If the early stages in cancer progression involve somatic mutations or chromosomal aberrations, impaired DNA repair efficiency could explain why families with increased predisposition move quickly through the early stages. When they have progressed through the early stages, individuals from those families have a high constant risk later in life while awaiting the final transition. By contrast, better repair efficiency slows the transition through the early stages. Slow transitions early in life mean more stages to pass through later in life. With more stages remaining, individuals at low risk continue to show an increase in incidence with age (Frank 2004d).

11.2 Progression and Incidence Affect Genetic Variation

The previous section described how genetic variants affect progression and incidence: the pathway from genes through development to phenotype. In this section, I analyze how progression and incidence affect the frequency of variants in populations: the pathway from phenotype through natural selection to gene frequency.

EVOLUTIONARY FORCES

Many forces potentially influence gene frequency. The wide range of alternatives makes it easy to fit some model to the observed distribution of frequencies, but hard to determine if the fit has any meaning.

Only natural selection provides a simple comparative prediction: the stronger the deleterious effect of a cancer-predisposing variant on survival and reproduction, the lower the expected frequency of that variant. A comparative prediction forecasts the overall tendency or trend, not the relative frequency of any particular variant.

In this section, I summarize the major evolutionary forces. The following section evaluates the comparative prediction that the deleterious effects of a variant influence its frequency.

DRIFT

Drift encompasses various chance events. Each copy of a genetic variant lives an individual and descends, on average, to λ babies. Most populations neither grow nor shrink continually, and so the total number of gene copies remains about the same with $\lambda \approx 1$. If the population shrunk in one generation to 10% of its current size, then $\lambda = 0.1$.

A few simple calculations illustrate the key role of drift for rare variants. Consider a population of size N with a particular variant at frequency p. In one generation, how much does p typically change if random drift is the only evolutionary force acting?

The number of copies of a particular variant is $\alpha = p2N$, where N is the size of the population, and $2N$ is the total number of gene copies—the factor of 2 arises because each diploid individual carries two copies of each gene.

In the next generation, the number of variant gene copies follows a Poisson distribution with an average of $\alpha\lambda$ in a progeny gene pool of size $2N\lambda$. As long as $\alpha\lambda$ is not too small, we can use the normal approx-imation for the Poisson distribution, which tells us that the number of variant gene copies in the next generation approximately follows a nor-mal distribution with mean $\alpha\lambda$ and standard deviation $\sqrt{\alpha\lambda}$. In terms of variant gene frequency p in the next generation, the 95% confidence interval is $p(1 \pm 2/\sqrt{\alpha\lambda})$.

How much does drift change gene frequency in one generation in a stable population, $\lambda = 1$? Suppose the gene frequency starts at $p = 10^{-5}$ in a gene pool of size $2N = 10^7$, so there are originally $\alpha = p2N = 100$ variant gene copies. In the next generation, the frequency of the variant gene has a 95% confidence interval of $p(1 \pm 0.2)$, which shows that 5% of the time the gene frequency will change by more than 20% in one generation. Over relatively short time periods, significant changes in the frequency of rare variants may occur.

LINKAGE AND HITCHHIKING

Consider a new variant that exists as a single copy in the population at frequency $p = 1/2N$. Suppose that focal variant resides on a chromo-some near another site that has a rare, favorable variant. Let the only force acting on the focal variant be the benefit derived from residing near a favorable variant at a nearby site.

Suppose the neighboring site causes an average increase in reproduc-tion of $1 + s$ compared with the normal value of one. Further, suppose the focal site and beneficial neighbor recombine at a rate of r per gen-eration. Then the frequency of the focal site tends to increase if $s > r$, that is, if the selective benefit, s, of being linked to an advantageous al-lele is greater than the rate, r, at which that linkage is broken down by recombination. If the selective benefit happens to be fairly strong, then the beneficial site will significantly increase the frequency of all of the closely linked variants.

PLEIOTROPY

Many variants affect more than one phenotype or more than one com-ponent of survival and reproduction. Suppose, for example, that a vari-ant enhanced the rate of wound healing. On the one hand, rapid healing would probably provide some benefit, perhaps against infection. On the other hand, wound healing can be carcinogenic probably because of the enhanced rate of symmetric mitoses, and more rapid wound healing may be more carcinogenic. So a variant that increased the rate of wound healing might rise to high frequency even though it shifts cancer incidence to earlier ages.

In general, when a variant shifts cancer to earlier ages and occurs at unexpectedly high frequency, pleiotropy is a reasonable hypothesis. However, it is often difficult to figure out the multiple effects of a variant and the respective consequences for survival and reproduction.

OVERDOMINANCE AND EPISTASIS: VARIABLE GENETIC BACKGROUND

Overdominance occurs when, at a locus with two alternative alleles, the heterozygote is more fit than either homozygote. Sickle cell anemia provides the classic example. An individual with one sickle cell variant allele enjoys protection against malaria, but an individual with two copies of the variant suffers severe disease from aberrations in red blood cells. Those opposing benefits and costs influence the frequency of the sickle cell variant.

Overdominance probably occurs rarely for variants that directly cause significant shifts of cancer to earlier ages. Most carcinogenic variants act in a physiologically recessive way, such that a cell with one normal copy and one variant copy has a normal phenotype. Deleterious effects at the cellular level arise only when both allelic copies suffer loss of function. However, an individual needs to carry only one mutated copy to be at risk; the cancerous phenotype arises after somatic mutation knocks out the second copy in a small fraction of cells. So, although most cancer-predisposing mutations are physiologically recessive, they are inherited as dominant alleles (Marsh and Zori 2002). So far, only three genes (*RET*, *MET*, and *CDK4*) have been found with inherited variants that act dominantly within cells (as oncogenes) among 31 cancer genes with single locus predisposing variants (Marsh and Zori 2002).

Pleiotropic overdominance may occur, in which a heterozygote locus that predisposes to cancer has beneficial effects on some other phenotype. Probably some cases of pleiotropic overdominance will eventually be discovered, but no evidence presently suggests this process as a major force maintaining genetic variability in predisposition.

Epistasis arises when the effect of a variant depends on the presence or absence of variants at other loci. Epistasis is much like overdominance: both processes cause changes in the phenotypic consequences of a variant in relation to the genetic background in which the variant lives. One can think of copies of the variant as living in genetically variable environments, favored in some environments and disfavored in others.

VARIABLE ENVIRONMENT

External environments also vary. For example, a variant may be disfavored in certain carcinogenic environments and favored in the absence of those environments. The variable selection can maintain variants that predispose to cancer at frequencies higher than expected through the deleterious effects of increased cancer incidence.

MUTATION AND SELECTION

When thinking about cancer, we can often take a simple point of view: mutation creates deleterious variants that predispose to cancer, and selection removes those deleterious variants from the population. The other evolutionary forces listed above may or may not act in any particular case, but deleterious mutation and the purging of those mutations by natural selection occur continually. The balance between mutation and selection sets the default against which we should compare observed frequencies.

Mutation-Selection Balance: A Comparative Prediction

It is often difficult to measure precisely the rate of mutation and the rate at which natural selection purges deleterious mutations. In addition, other forces such as drift and pleiotropy often affect the frequency of deleterious, predisposing variants. So any attempt to predict precisely the frequency of a deleterious variant or to fit some model with estimated parameters of mutation and selection would mislead: one can calculate precise predictions or estimate parameters, but those calculations or estimations would only provide a false sense of precision.

We can estimate the relative strengths of mutation and selection within an order of magnitude or so. Those rough estimates provide guidelines to the expected frequencies of deleterious variants. We can also make two simple comparative predictions. First, as selection against variants increases, the observed frequency of the variants declines. Second, as mutation rate at a particular locus increases, the observed frequency of deleterious variants at that locus increases.

These rough guidelines and comparative predictions set a baseline for expectations of variant allele frequency. When observations deviate significantly from expectations, then we may turn to forces other than a balance between deleterious mutation and purging by natural selection.

HIGH PENETRANCE AND EARLY ONSET

Suppose a mutation is expressed in all carriers, and those carriers die before they have reproduced. In this situation, each case must arise from a new mutation, and the frequency of mutated alleles, q, is roughly equivalent to the mutation rate per generation, u, that is, $q = u$.

Inherited cases of retinoblastoma, Wilms' tumor, and skin cancer in xeroderma pigmentosum transmit as dominant mutations. Most individuals who carry a highly penetrant mutation develop the disease during childhood or early life. Without treatment, carriers do not usually reproduce. These diseases all occur at frequencies, q, of approximately 10^{-5}-10^{-4} (Vogelstein and Kinzler 2002).

The commonly quoted values for mutation rate, u, tend to be in the range of 10^{-6}-10^{-5} per gene per generation (Drake et al. 1998), an order of magnitude lower than the frequency of cases. For this type of approximate calculation, a match within an order of magnitude suggests that we have roughly the right idea about the factors that influence allele frequencies.

Certainly, other estimates of frequency for these diseases or other early-onset cancers will not match so closely to the usual estimate of the mutation rate. A mismatch implies some force beyond the standard baseline mutation rate and immediate removal of all mutations by natural selection. For example, the penetrance may be less than perfect, some carriers may reproduce, or the gene may be unusually mutable.

AGE OF ONSET AND THE FORCE OF SELECTION

Some inherited mutations have low penetrance or cause later-onset disease. Natural selection removes a mutation from the population in proportion both to the probability that it causes disease and to the reduction in reproductive success of those individuals who express the disease (Rose 1991; Nunney 1999, 2003; Frank 2004e). Reduction in reproductive success depends on the age of onset: later onset has less effect on transmission of alleles to the next generation. Figure 11.8

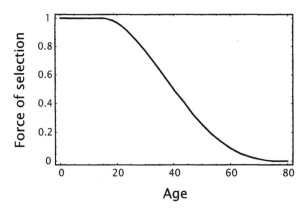

Figure 11.8   The force of selection at different ages. Loss in fitness caused by
cancer is the force of selection averaged over the probabilities of death at different ages. This loss is $pr = \int_0^T \dot{x}_n(t)f(t)dt$, where pr, the fractional loss in
fitness, is the averaged product of the age-specific incidence, $\dot{x}_n(t)$, and the loss
in reproduction caused by death at age t, $f(t)$. The age-specific incidence provides a measure of penetrance at different ages. No good data exist to estimate
the force of selection at different ages for humans; however, the curve shown
here gives the approximate shape of the force of selection.

shows the technical details. The following paragraphs describe the main
points.

   Suppose the probability of expression in a carrier—the penetrance—
is p, and the reduction in reproductive success is r. If q is the frequency
of the mutant allele in the population, then qp is the frequency of cases,
and the rate at which mutations are removed in each generation is qpr,
the frequency of cases multiplied by the reduction in reproductive success in each case. Equilibrium occurs when mutant alleles purged by
selection match the influx of new mutations at rate u, so at equilibrium,
$qpr \approx u$.

Familial adenomatous polyposis.—Inherited mutations of the *APC* gene
act in a dominant manner and cause the colon cancer syndrome familial
adenomatous polyposis (FAP) (Kinzler and Vogelstein 2002). Nearly all
carriers develop cancer, with a median age of onset of about 40 years.
The frequency of cases, qp, is of the order of 10^{-4}. We do not have
historical data on the reduction in reproductive success that occurs in
the absence of treatment. A reasonable value is $r \approx 10^{-1}$, which takes

into account the fact that the age of reproduction in the past was probably somewhat lower than in modern societies. In this case, $qpr \approx 10^{-5}$,
which is again fairly close to the standard estimate for the mutation rate.

Hereditary nonpolyposis colon cancer.—Mutations in the DNA mismatch
repair (MMR) system lead to hereditary nonpolyposis colon cancer (HN-
PCC) (Boland 2002). Mutations in several MMR genes cause an increase
in the somatic mutation rate, and more frequent somatic mutations lead
to a high probability of early-onset cancer. The median age of diagnosis
for HNPCC is about 42 years (Lynch et al. 1995). The frequency of cases
is at least of the order of 10^{-3}, but may be more because HNPCC can be
difficult to distinguish from colon cancers that arise in the absence of
MMR defects.

Setting the level of reproductive loss at $r = 10^{-1}$, the rate of removal
of MMR mutations, qpr, is 10^{-4} or higher. This value would indicate a
high mutation rate if there were only one MMR locus. However, mutations that increase the risk of developing HNPCC have been identified in
five MMR loci so far (Boland 2002), and mutations that influence HNPCC
may also occur in other MMR genes. There are 22 genes in the core MMR
pathway (Mohrenweiser et al. 2003). The effective mutation rate is nu,
where n is the number of MMR loci and u is the mutation rate per locus.
Using a range for n of approximately 3–10, we obtain a range for the
mutation rate per locus of approximately $1–3 \times 10^{-5}$.

Neurofibromatosis type 1.—Inherited mutations in the neurofibromatosis 1 (*NF1*) gene cause a variety of symptoms with variable penetrance
(Gutmann and Collins 2002). Carriers may express various nonlethal
deformities: numerous flat, pigmented skin spots; freckling; pigmented
nodules of the iris; and soft, fleshy peripheral tumors that arise from
nerves (neurofibromas). Several other complications develop, including
seizures, learning disabilities, and scoliosis.

NF1 is among the most common dominantly inherited diseases of
humans. Gutmann and Collins (2002) estimated prevalence of about
3×10^{-4}, based on several earlier studies (Crowe et al. 1956; Huson et al.
1989; Sergeyev 1975; Samuelsson and Axelsson 1981). Carriers almost
always express some of the symptoms—a penetrance, p, of nearly one.
The disease rarely reduces potential fertility, but actual reproductive
success of carriers has been estimated to be about one-half of normal

individuals, $r \approx 0.5$ (Huson et al. 1989). Thus, $qpr \approx 10^{-4}$, which implies a high germline mutation rate.

Few families transmit a mutation through several generations, and most cases arise from new mutations (Gutmann and Collins 2002). A wide variety of DNA lesions occur in the gene, including translocations, large chromosomal deletions, smaller deletions within the gene, small rearrangements within the gene, and point mutations. No particular mutational hotspots have been detected. This large gene spans almost 9 kb of coding DNA over at least 57 exons and, including the intron regions, approximately 300 kb of total DNA. Perhaps the large size contributes to the high rate at which loss of function mutations arise. It will be interesting to learn if other special attributes of this gene cause the apparently elevated mutation rate.

Hereditary breast cancer.—Mutations in *BRCA1*, which has an important function in the repair of double-strand DNA breaks, confer a high probability of developing breast or ovarian cancer (Couch and Weber 2002). Current estimates for the penetrance of breast cancer in carriers of *BRCA1* mutations range from 56% to 86% (Couch and Weber 2002). Lack of functional BRCA1 leads to chromosomal abnormalities (Welcsh and King 2001), a common feature of cancer cells. The median age of onset is approximately 50 years (Ford et al. 1998), which is later than for most of the other cancers that follow dominant Mendelian inheritance. The frequency of *BRCA1* mutant alleles and associated cases varies in different populations over the range 10^{-3}-10^{-2} (Tonin et al. 1995; Couch and Weber 1996; Struewing et al. 1997; Couch and Weber 2002). No data measure the decrease in reproduction in carriers of *BRCA1* mutations: a reasonable guess would be in the range 10^{-2}-10^{-1}. These values give an estimate for qpr of 10^{-5}-10^{-3}, which is somewhat higher than the standard assumption of 10^{-6}-10^{-5} for the mutation rate.

Welcsh and King (2001) suggested that *BRCA1* may have an elevated somatic mutation rate because of the high density of repetitive DNA elements in the gene. Those repeats may also cause a higher germline mutation rate, which would explain the higher than expected frequency of variants in populations.

Alternatively, Harpending and Cochran (2006) argued that natural selection of *BRCA1* variants may be more strongly affected by that gene's role in early brain growth and development rather than in DNA repair.

Such pleiotropy could explain the elevated frequency of *BRCA1* if the variants had beneficial effects on brain development. In particular, Harpending and Cochran (2006) argue that heterozygotes for *BRCA1* variants can in some environments have beneficial neural effects, but the variant homozygotes would be at a disadvantage. A mild heterozygote advantage balanced against strongly deleterious effects in the variant homozygotes could explain the observed frequency of *BRCA1* variants. The age of variant *BRCA1* alleles may provide clues about the forces that affect allele frequencies.

THE AGE OF ALLELES: A COMPARATIVE PREDICTION

Variants that cause greater reproductive loss will disappear from the population faster than variants that cause relatively lower reproductive loss.

In the simplest case, each new variant causes early death before reproduction, and each variant only lives for a generation. Lower penetrance or later onset imposes a weaker selective sieve against variants, allowing the variants a longer time before extinction.

Soon, we will have enough data on the DNA sequences of variants to allow reconstruction of their history and the time back to their common ancestor—the age of the allele. If the age of alleles is primarily determined by a balance between the origin of novel variants by mutation and clearance from the population by selection, then those ages should follow the simple prediction that more deleterious alleles tend to last a shorter period of time. Alternatively, forces other than mutation-selection balance may determine the age of alleles.

Consider, for example, the two alternative hypotheses for *BRCA1* variant frequency. If the elevated frequency of *BRCA1* variants arises from a higher germline mutation rate for that gene balanced against continual loss of variants by selection, then most variants at this locus should be relatively young (recent in origin). By contrast, if the elevated frequency arises from pleiotropic beneficial effects on neural development balanced against deleterious effects on cancer progression, then most variants at this locus should be relatively old.

11.3 Few Common or Many Rare Variants?

I have discussed a small number of mutations in which carriers suffer significantly earlier onset of disease. In those cases, a single mutation greatly increases incidence. Such mutations often appear to occur in key genes that directly affect progression of the particular type of cancer.

The search for single mutations of large effect has intensified over the past few years. However, few new mutations have been discovered. Most of the inherited predisposition to cancer remains unexplained. The widespread heritability of cancer appears to be caused by several variants each of relatively small effect—what is often called *polygenic* inheritance.

Within this large, polygenic component of heritability, do genetic variants that cause disease tend to be common or rare? Are there relatively few common, older variants or many rare, newer variants?

Much recent debate in biomedical genetics has turned on these questions, because methods for estimating genetic risk in particular individuals depend on the frequency of variant alleles (Weiss and Terwilliger 2000; Lee 2002). If most genetic risk comes from a few relatively common alleles that are relatively old, then those alleles will be associated with other polymorphisms in the genome that can be used as markers of risk. Those associations arise because the original mutations will, by chance, occur in regions in which other single nucleotide polymorphisms (SNPs) are located nearby.

By contrast, most genetic risk might come from many rare, young alleles. If so, then there will be no consistent association between known SNPs and genetic predisposition. Each particular mutation will have its own profile of linked marker polymorphisms, often specific for a particular population. Those linkage profiles will differ for each mutation. Because there may be many mutations, with each making only a small contribution to genetic risk, no overall association will occur between known marker polymorphisms and total genetic risk.

The available data do not definitively distinguish between a few common, older variants and many rare, younger variants. Wright et al. (2003) argued eloquently in favor of many rare variants; I agree with their logic. However, the issue here does not turn on point of view, but rather on the actual distribution of variants and their effects. I discuss two examples that provide the first clues.

Multiple Colon Adenomas

Fearnhead et al. (2004) collated data on 124 individuals with multiple adenomatous polyps. They screened those individuals for germline DNA variants in five genes known to influence colon cancer progression, and found 13 different variants. They compared the frequency of those 13 variants in the 124 cases with the frequency in 483 random control individuals.

Table 11.1 shows the frequencies of the 13 variants in cases and controls. These results suggest that many rare variants, each of small effect, contribute significantly to the heritability of cancer. In this study, almost all of the variants were single amino acid substitutions. Each such small change in protein shape and charge may contribute a small amount to disease. Many such changes, each rare, may in the aggregate explain much of the genetic basis of disease.

Fearnhead et al. (2004) support their argument that single amino acid substitutions in proteins contribute to disease by evaluating the functional changes for many of the mutations listed in Table 11.1. Almost all of the variants occur in regions of their proteins known to have important functional roles in pathways that are often disrupted in tumors. I briefly summarize two examples from Fearnhead et al.'s (2004) discussion.

The *APC* variant *E1317Q* alters charge in the region that binds to β-catenin. Mutation of the APC regulatory pathway appears to be a common first step in adenoma formation (Kinzler and Vogelstein 2002). APC represses β-catenin, which may have two different consequences for cellular growth. First, β-catenin may enhance expression of c-Myc and other proteins that promote cellular division. Second, β-catenin may play a role in cell adhesion processes, effectively increasing the stickiness of surface epithelial cells. In either case, repression of β-catenin reduces the tendency for abnormal tissue expansion. In tumors, somatic mutations in *APC* usually include domains involved in binding β-catenin, releasing β-catenin from the suppressive effects of APC (Kinzler and Vogelstein 2002).

The *hMLH1* variant *K618A* alters the charge of a highly conserved region of this DNA mismatch repair protein. Several deleterious mutations have been reported in this region (Wijnen et al. 1996; Peltomaki and Vasen 1997; Mitchell et al. 2002), and studies in yeast demonstrated

Table 11.1   Variants in cases with multiple polyps and in controls

Gene	Mutation	% carriers in cases	% carriers in controls
APC	*E1317Q*	2.42	1.25
CTNNB1	*N287S*	0.81	0.62
AXIN1	*P312T*	0.81	0.00
AXIN1	*R398H*	0.81	0.00
AXIN1	*L445M*	0.81	0.00
AXIN1	*D545E*	1.61	1.28
AXIN1	*G700S*	4.84	3.96
AXIN1	*R891Q*	3.91	2.93
hMLH1	*G22A*	0.81	0.21
hMLH1	*K618A*	3.22	2.07
hMLH2	*H46Q*	0.81	0.00
hMLH2	*ex4SDS*	0.81	0.00
hMLH2	*E808X*	0.81	0.00
Combined		24.9	11.5

From Tables 2 and 3 of Fearnhead et al. (2004), who contributed new data and also collated data from various sources (Frayling et al. 1998; Lamlum et al. 2000; Webster et al. 2000; Dahmen et al. 2001; Taniguchi et al. 2002; Guerrette et al. 1998; Tannergard et al. 1995). The mutation *ex4SDS* is an exon 4 splice donor site. Mutations of the form $\alpha\#\beta$ describe amino acid substitutions $\alpha \to \beta$ at codon position #.

that substitutions at position 618 cause functional changes (Shimodaira et al. 1998). hMLH1 works in various heteromeric complexes, including interaction with hPMS2 (Buermeyer et al. 1999; Fishel 2001); the *hMLH1 K618A* mutation causes more than 85% loss of interaction between hMLH1 and hPMS2 (Guerrette et al. 1999).

DNA REPAIR VARIANTS

Earlier in this chapter, I mentioned that DNA repair efficiency varies considerably in populations and has a large heritable component (Grossman et al. 1999; Cloos et al. 1999; Roberts et al. 1999). In addition, poor repair efficiency consistently associates with an approximately 2–10-fold increase in cancer risk (Berwick and Vineis 2000).

The previous section showed that rare variants at DNA mismatch repair loci can predispose to colon cancer. The fact that rare variants can predispose does not resolve whether the high heritability of repair efficiency and cancer predisposition arises mainly from relatively rare or common alleles. The existing data do not settle the issue. Two lines of evidence provide clues.

FREQUENCY DISTRIBUTION OF VARIANTS

Mohrenweiser et al. (2003) summarized genetic variation across 74 DNA repair loci. Figure 11.9 shows that the rare, intermediate, and common alleles contribute equally to the variance in allele frequency. To understand what this means, consider how to calculate the genetic variance in allele frequencies.

The contribution of a variant allele with frequency p_i to the variance at its locus is $v_i = p_i(1 - p_i)$. A rare allele at frequency $p_i = 0.01$ contributes $v_i \approx 0.01$ to the frequency variance. A common allele at frequency $p_i = 0.11$ contributes $v_i \approx 0.1$ to the frequency variance, or about an order of magnitude more than the rare variant. If there were ten times as many rare variants as common variants, then the rare and common variants would contribute equally to the total variance.

Figure 11.9 shows that there are more rare variants than common variants. The excess of rare variants explains why the total contribution to the variance in allele frequency is about the same for rare, intermediate, and common alleles.

These calculations provide information about the frequency of variant alleles. However, these data do not connect the different variants to their consequences for disease. Inevitably, some of the variants will have little or no effect, whereas others may significantly increase risk. The common types are unlikely to be severely deleterious, but beyond that, no strong conclusions can be made about the effects of the variant alleles. The data on colon cancer in the previous section show that rare variants can influence predisposition. The next section shows that combinations of common variants may also significantly affect predisposition.

MULTIPLE VARIANTS INCREASE PREDISPOSITION

A pathway such as a particular type of DNA repair forms a quantitative trait that protects against cancer progression. Certain individual polymorphisms may each reduce the efficacy of the pathway by a small amount, and consequently cause a small and perhaps undetectable increase in cancer risk. In combination, multiple polymorphisms may significantly reduce efficacy and consequently cause a significant rise in

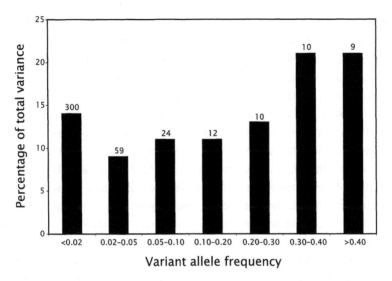

Figure 11.9   The relative variance in allele frequencies for rare and common alleles of 74 DNA repair genes. The total number of variants in each frequency category is shown above the bars. Each rare variant contributes a small fraction of the total variance, but there are many more rare than common variants. Changes in amino acid sequence define the variants. Data collated by Mohrenweiser et al. (2003).

cancer risk. Particularly high risk may occur when those polymorphisms concentrate in one or more key pathways and compromise essential protective mechanisms (Han et al. 2004; Popanda et al. 2004; Cheng et al. 2005; Gu et al. 2005; Wu et al. 2006).

Wu et al. (2006) measured the frequency of 44 polymorphisms in variant DNA repair and cell-cycle control genes. They compared frequencies in 696 patients with bladder cancer versus 629 unaffected controls. The study focused on the increase in relative risk with a rise in the number of variant alleles. The hypothesis was that many cases would arise in individuals who carry a greater than average number of predisposing polymorphisms in key pathways.

To analyze the role of multiple variants in a sample of modest size, one must study relatively common variants. If the variants were rare, very few individuals would carry several variants. Thus, the design of Wu et al.'s (2006) study focuses attention on the role of multiple common variants, without addressing how multiple rare variants may contribute to disease. In spite of this limitation, the study is important because much of polygenic predisposition may arise from the combined effect of many variants. Given the widespread distribution of variant alleles in populations (Figure 11.9), each individual carries a unique combination of numerous variants across key pathways in carcinogenesis.

Wu et al.'s (2006) most interesting result concerns the interaction between smoking and polymorphisms in the DNA repair pathway that functions in nucleotide-excision repair (NER). The NER pathway removes bulky DNA adducts frequently caused by the polycyclic aromatic hydrocarbons in tobacco smoke. Smoking significantly increases bladder cancer risk. A few studies have shown that certain single polymorphisms within the NER pathway associate weakly with greater susceptibility to bladder cancer (reviewed by Garcia-Closas et al. 2006). Such weak effects are often difficult to reproduce in subsequent studies.

Wu et al. (2006) included 13 NER variants across nine loci. Among those who have smoked, individuals with seven or more NER variants had a relative risk of cancer 3.37 times greater than those with fewer than four variants, with a 95% confidence interval for relative risk of 2.08–5.48. Among nonsmokers, individuals with seven or more variants had a relative risk of cancer 1.40 times greater than those with fewer than four variants, with a 95% confidence interval for relative risk of 0.72–2.73.

Wu et al. (2006) further analyzed all 44 polymorphisms across 33 DNA repair and cell-cycle control loci. Among the 851 individuals who had smoked, 74% of the subjects had bladder cancer. The most powerful genetic effect concentrated in the NER loci: among the 124 smokers who carried three particular NER variants, 97% had bladder cancer, whereas only 53% of those smokers who did not carry all three variants had bladder cancer.

The results in Wu et al.'s (2006) study suggest that multiple NER variants significantly raise cancer risk in smokers. Such studies are often difficult replicate for at least three reasons.

First, the strong effect of smoking demonstrates that certain polymorphisms may only have strong effects in the presence of particular environmental challenges. Unmeasured environmental or genetic effects may often determine whether the particular genotypes under study play an important role in progression.

Second, the variants under study may not directly affect progression, but instead be linked to variants at other sites that influence carcinogenesis. In other populations, with different genetic linkage relations, those same variants will associate differently with cancer rates.

Third, such studies suffer from problems common to exploratory statistical analyses: the number of variables (polymorphisms) and their combinations greatly exceeds the number of individuals sampled. With so many different combinations, by chance certain combinations will associate with strong differences in outcome. Although statistical methods attempt to deal with such problems, conclusions from such studies often do not hold up in future attempts to repeat the work.

With those caveats in mind, I now compare Wu et al.'s (2006) results with a similar study. Garcia-Closas et al. (2006) analyzed 22 polymorphisms in seven NER genes among 1,150 bladder cancer cases and 1,149 controls. In agreement with Wu et al. (2006), Garcia-Closas et al. (2006) found weak effects for each variant when analyzed in isolation, but found stronger, significant effects when analyzing the interaction between smoking and multiple NER variant sites. Garcia-Closas et al. (2006) limited their analysis to pairs of variant NER sites, and found that certain pairs of variants significantly increased risk in smokers.

The two studies had six NER polymorphisms in common. Four of those polymorphisms were not particularly important in either study. At the locus *RAD23*, one particular variant played a key role in Wu et al. (2006) but, although present in Garcia-Closas et al. (2006), did not play a key role in that study. Instead, Garcia-Closas et al. (2006) found that a different variant site in *RAD23* had significant explanatory power when evaluating interactions between pairs of variant sites. The two studies also shared a variant at the *ERCC6* locus: that variant was important in multisite interactions in Wu et al. (2006) but not in Garcia-Closas et al. (2006).

CONCLUSION

Preliminary evidence suggests that risk depends on the combination of effects at multiple variant sites. Practical sampling issues limit studies to combinations of common variants. In small samples, combinations of rare variants occur too infrequently to allow study. In the population, more rare variants occur than common variants (Figure 11.9), so the net contribution of multiple rare variants may be at least as great as the combinations of common variants.

The effect per variant of rare versus common variants remains unknown. Rare alleles will likely have greater effects than the common alleles if variant frequency depends on mutation, drift, and selection against deleterious effects. By contrast, common alleles may have larger effects if variants either have variable consequences depending on environment or genetic background, or if variants have beneficial pleiotropic effects that offset the deleterious traits that increase cancer incidence.

It will not be easy to work out the relative contribution of different variants and how variants combine to determine disease. But much attention will continue to focus on this problem. Through cancer studies, we will gain insight into the genetic basis of variability in key functional components, such as DNA repair and tissue regulation via control of the cell cycle. By study of functional components and their genetic basis of variation in efficiency, and how the components interact to determine disease, we will begin to understand how evolution has shaped the age-specific curves of failure. Through those curves of failure, we can analyze the evolutionary design of reliability that sets the nature of disease and aging.

11.4 Summary

The first part of this chapter described how inherited genetic variants affect the age of cancer onset. In the future, new genomic technologies will measure genetic variation with far greater resolution. To interpret those high-resolution measurements of genetic variation, we will have to connect the observed genetic variation to the causes of cancer. Such connections can only be made by studying how genetic variants shift the age-specific incidence. In the second part of the chapter, I analyzed the population frequency of predisposing genetic variants in light of various evolutionary forces. I suggested that studies of cancer predisposition may lead the way in understanding the structure of inherited genetic variation for age-specific diseases.

The next chapter turns to the somatic evolution of cancer within individuals. Most human cancers arise in tissues that renew throughout life. Those tissues often derive from stem cells. I review the biology of stem cells and how the shape of stem cell lineages in renewing tissues affects the progression of cancer.

12 Stem Cells: Tissue Renewal

Tissue renewal determines the rate of cell division. In many tissues, renewal derives from rare stem cells. In this chapter, I discuss how mitotic rate and lineal descent from stem cells set the relative risk of cancer.

The first section provides background on tissue renewal and cancer. About 90% of human cancers arise in epithelial tissues. Epithelial layers in certain organs, such as the intestine and skin, renew continuously throughout life. Cancer incidence in renewing tissues rises sharply with age. By contrast, childhood cancers concentrate in tissues that divide rapidly early in life but relatively little later in life. In general, the age-specific rate of cell division explains part of the relative risk for different tissues at different ages.

The second section describes the shape of cell lineages in renewing tissues. Many tissues that renew frequently have a clear hierarchy of cell division and differentiation. Rare stem cells divide occasionally, each division giving rise on average to one replacement stem cell for future renewal and to one transit cell. The transit cell undergoes multiple rounds of division to produce the various short-lived, differentiated cells. New stem cell divisions continually replace the lost transit cells. I review the stem-transit architecture of cell lineages in blood formation (hematopoiesis), in gastrointestinal and epidermal renewal, and in sex-specific tissues such as the sperm, breast, and prostate.

The third section discusses the important distinction between symmetric and asymmetric stem cell divisions. In symmetric divisions, the two daughter cells have an equal chance to remain a stem cell or differentiate into a transit cell. To maintain a pool of N stem cells in a niche, each stem cell division produces on average one new stem cell and one new transit cell; the fate of each cell is determined randomly. In asymmetric divisions, differentiation happens in a determined way: one particular daughter cell remains a stem cell, and the other differentiates into a transit cell.

The fourth section analyzes how symmetric versus asymmetric stem cell divisions affect the accumulation of mutations over time. In every

mitosis, the DNA duplex splits, each strand acting as a template for replication to produce a new complementary strand. Most mutations during
replication probably arise on the newly synthesized strand. Under a
program of asymmetric cell division, a stem lineage could reduce its
mutation rate if each stem cell division segregated the oldest template
strands to the daughter destined to remain in the stem lineage and the
newer strands to the daughter destined for the short-lived transit lineage. Recent evidence supports this hypothesis of strand segregation in
stem cell lineages.

The fifth section outlines how tissue compartments prevent competition between cellular lineages. In tissues such as the intestine and
skin, the spatial architecture restricts lineal descendants of stem cells
to a very narrow region. From a lineage perspective, each compartment
limits the local population size and defines a separate parallel line of
descent and evolution. An expanding clone, perhaps one step along in
carcinogenesis, cannot normally grow beyond its compartmental boundaries, thus limiting the target number of cells for the accumulation of
subsequent mutations.

12.1 Background

TISSUE DEMOGRAPHY AND THE DISTRIBUTION OF TUMORS

Roughly 90% of cancers arise as carcinomas in epithelial (surface) tissues. The epithelium may be the external surface of an organ, such as
the skin or outer lining of the intestine, or internal surfaces of the bladder, prostate, breast, and so on. The other 10% of cancers arise mostly
as leukemias (blood) and sarcomas (connective tissues, bone, etc.).

Cairns (1975) listed the tissue distributions from the Danish Cancer
Registry, as shown in Table 12.1. Peto (1977) estimated that for fatal
cancers in Britain, 20% derive from sex-specific epithelial cells (breast,
prostate, ovary), 70% derive from other epithelial cells (lung, intestine,
skin, bladder, pancreas, etc.), and 10% derive from non-epithelial cells
(blood, bone, connective tissues, etc.).

The age-specific rate of cell division explains part of the relative risk
for different tissues. Rare childhood cancers concentrate in tissues that
undergo cell division early in life followed by relative cellular quiescence
(see Section 2.3). Common adult-onset cancers occur in surface epithelia
that renew throughout life, such as in the skin and intestine.

Table 12.1   Cancer incidence in Denmark, 1943–1967

Type	Commonest sites	Total cases	%
Carcinomas			
External epithelia	Skin, large intestine, lung, stomach, cervix	168,591	56
Internal epithelia	Breast, prostate, ovary, bladder, pancreas	110,182	36
Sarcomas and leukemias		23,801	8

From Cairns (1975), based on data from the Danish Cancer Registry (Clemmesen 1964, 1969, 1974).

RENEWING TISSUES AND EPITHELIAL RISK

The epithelium of the human colon turns over at least once per week throughout life. As cells die at the surface, they are replaced by new cell divisions. By age 60, a person has been through at least 3,000 replacement cycles, which means that some cell lineages must pass through many generations. Those renewing lineages would be at high risk for accumulating mutations and progressing to cancer.

Cairns (1975) recognized the importance of tissue renewal in the distribution of cell divisions, and the key role that cell division plays in cancer progression. He wrote:

> We may ... expect to find, especially in animals which undergo continual cell multiplication during their adult life, the evolution of mechanisms that protect the animal from being taken over by any "fitter" cells arising spontaneously during its lifetime—that is mechanisms for minimising the rate of production of variant cells and for preventing free competition between cells ... Because most of the cell division is occurring in epithelia, that is where we may expect to find the protective mechanisms most highly developed.

12.2 Stem-Transit Program of Renewal

Cairns (1975) suggested various mechanisms that protect against the accumulation of somatic mutations and the competition between cell lineages.

One protective mechanism arises from the distinction between stem cells and transit cells. The long-lived stem cells renew the tissue over

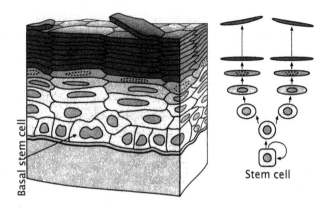

Figure 12.1   Pattern of cell division in the epithelial layer of the skin. At the deepest layer, each basal stem cell divides and produces one cell that remains at the base to continue as a stem cell and one cell that moves up to form the transit lineage. The transit lineage divides a few times, and the cells progress through various developmental stages as they migrate to the surface. Eventually, the cells lose their nucleus and synthesize the insoluble proteins of the skin (keratin). As the basal stem cells continue to divide, the flow of cells from the basal layer pushes the cells above toward the surface. The surface layer continually sheds dead cells, which are replaced by new cells from below. From Figure 4.1 of Cairns (1997).

many years. The short-lived transit cells derive from stem cells, divide several times to provide a temporary population of surface cells, and then die. Cairns (1975) wrote:

> The turnover that occurs in the self-renewing epithelia is the result of continual shedding of superficial cells balanced by continual multiplication of the deeper cells. In the simplest examples, like the skin, cell division is restricted to the deepest (basal) layer of cells [Figure 12.1]. To keep the number of basal cells constant, one of the two daughter cells resulting from each cell division must on average remain in the basal layer and the other must escape and be discarded.

Cairns contrasted two alternative patterns by which tissues may renew themselves. In Figure 12.2a, the lower left cell is the single stem cell that will renew the local area of tissue. Each stem cell division produces one new daughter stem cell to the right and one new transit cell to the top. The transit cell migrates up through the tissue and dies on the surface. The new stem cell repeats the process. Through 16 cell divisions, the original stem cell produces 16 new transit cells that renew

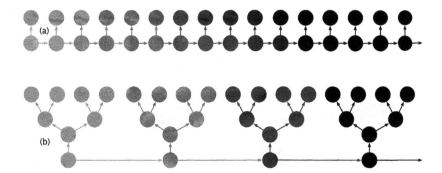

Figure 12.2   Alternative stem-transit designs to renew a tissue based on asymmetric stem cell division. Each pattern begins with a single stem cell at the lower left. Time moves to the right, as the stem lineage progresses along the lower row in each case. Stem cells divide asymmetrically in these two patterns, each stem cell division producing one daughter transit cell and one daughter stem cell. All cells that remain in the tissue over time trace their ancestry back through a linear history of stem cell divisions. Derived from Cairns (1975).

the tissue over time. Those 16 stem cell divisions also trace a linear history of descent, so that the final stem cell on the bottom right traces its ancestry back through the lineage that forms the bottom row. Any mutations that remain in the tissue over time must occur in the stem cell lineage.

Figure 12.2b presents a second pattern by which the stem lineage may produce 16 transit cells. The original stem cell at the bottom left divides to produce one new daughter cell to the right and one new transit cell to the top. The transit cell then goes through two further rounds of cell division, producing four transit cells to renew the tissue for each stem cell division. In this case, the tissue produces 16 transit cells with just four rounds of stem cell division. Again, any mutations that remain in the tissue over time must occur in the stem cell lineage, but with just four stem cell divisions in (b), that pattern reduces the accumulation of mutations relative to the pattern in (a) with 16 stem cell divisions.

Those tissues that renew most often appear to have a stem-transit architecture, following the pattern in Figure 12.2b.

HEMATOPOIETIC RENEWAL

The numerous distinct blood cell types derive from hematopoietic stem cells via a complex transit hierarchy (Weissman 2000; Kondo et al.

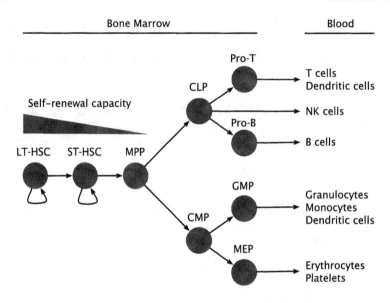

Figure 12.3   The transit lineage of hematopoietic differentiation in adult mice. Long-term hematopoietic stem cells (LT-HSC) renew throughout life. Short-term hematopoietic stem cells (ST-HSC) self-renew over a 6–8 week period. Multipotential progenitor (MPP) cells self-renew for less than two weeks, differentiating into common lymphoid progenitors (CLP) and common myeloid progenitors (CMP). Those progenitors then differentiate into another layer of precursors, which then differentiate into the final cell types of the blood. Redrawn from Kondo et al. (2003) and Shizuru et al. (2005).

2003). Figure 12.3 shows the differentiation hierarchy. Only the long-term (basal) stem cell lineage survives over time. The other cell lineages divide a limited number of times, differentiate, and die, to be replaced by new daughter cells derived from the basal stem lineage. I could not find any clear statement about the typical number of cell divisions from the basal lineage to extinction of a transit lineage.

The long-term stem cells of young mice appear to divide roughly every 10–20 days. No evidence suggests different rates of division between stem cells (Bradford et al. 1997; Cheshier et al. 1999).

GASTROINTESTINAL RENEWAL

Studies of mice and humans show that the epithelial surface of the intestine sloughs off continually and is renewed by fresh cells (Bach et al.

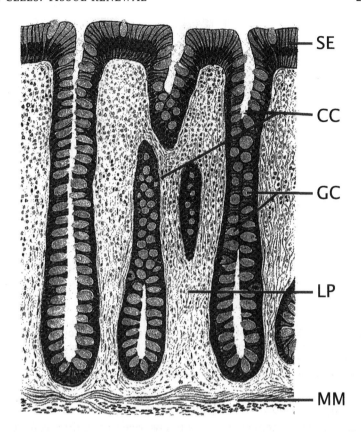

Figure 12.4   The morphology of normal colon tissue. Labels show surface epithelium (SE), colon crypts (CC), goblet cells (GC), lamina propria (LP), and muscularis mucosa (MM). The crypts open to the surface epithelium—in this cross
section, some of the crypts appear partially or below the surface. From Kinzler
and Vogelstein (2002), original published in Clara et al. (1974).

2000). Renewal occurs by a flow of cells from numerous invaginations—
crypts—throughout the intestinal surface (Figure 12.4). Cells flow from
the base of each long, narrow crypt to the surface.

   The small intestine of the mouse has about 15 cell layers from the epithelial surface to the base of the crypt (Figure 12.5). In the small intestine, stem cells reside around the fourth cell position from the bottom.
Those stem cells produce daughters that flow either down to the lowest
layers, where they differentiate into Paneth cells, or upward where the
daughter cells continue to divide and differentiate into the functional
goblet cells and enterocytes of the intestinal epithelium.

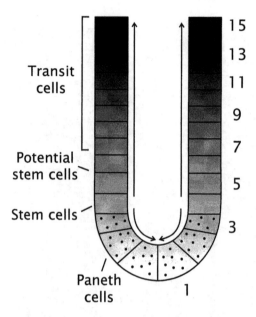

Figure 12.5   Schematic of a small intestine crypt of a mouse. The crypt has about 15 cells from the epithelial surface at the top to the base, as numbered along the right. In three dimensions, the cylindrical lining of the mouse small intestine crypt has about 200–250 cells. Modified from Marshman et al. (2002).

Figure 12.6 shows the cell lineage hierarchy of the mouse small intestine. The active stem cells divide to give rise to daughter cells. One-half of the daughter cells must remain active stem cells to continue future renewal. The other half of the daughters begins the transit pathway to differentiation.

In the first few transit divisions, T_1–T_3, the cells retain the potential to return to fully active stem cells in order to replace stem cells that die or to contribute to tissue renewal after injury. Some of those early transit lineage cells differentiate into Paneth cells and flow downward; the others continue to flow upward, divide, and eventually differentiate into the mature epithelial cells. Within a week or so, the daughters of the stem cells have flowed to the surface and died, to be replaced by the continual flow from below. Figure 12.7 gives a rough idea of the three-dimensional crypt architecture.

Gastrointestinal stem cells remain difficult to identify unambiguously. Through various indirect studies, Bach et al. (2000) conclude that each mouse small intestine crypt has 4–6 active stem cells. Those stem cells

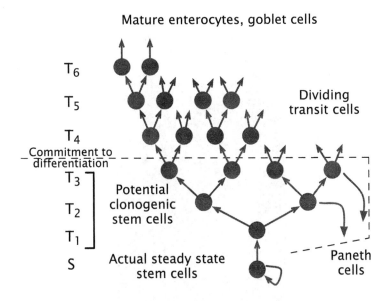

Figure 12.6 Cell lineage hierarchy in a small intestine crypt of a mouse. Active stem cells give rise to daughter stem cells that remain near the base of the crypt and the first generation of the transit lineage pathway (T_1), the potential clonogenic stem cells. The early transit generations retain the ability to return to fully active stem cells, but normally they move either up or down the crypt. If the cells move down, they differentiate into Paneth cells that line the base of the crypt. If the cells move up, they differentiate into goblet cells and then mature enterocytes, after which they die and are shed from the epithelial surface. Redrawn from Marshman et al. (2002).

divide about once per day; each crypt produces about 300 new cells per day. There are about six transit divisions, so it takes about one week for a daughter cell of the stem lineage to move up, differentiate, and die at the surface. The mouse small intestine has about 7×10^5 crypts, so the whole small intestine of the mouse produces about 2×10^8 cells per day.

The large intestine (colon) has a similar architecture but lacks Paneth cells. Cancer occurs more often in the large intestine than in the small intestine, in spite of the similar tissue architecture and pattern of cellular renewal. Probably the colon suffers greater concentrations of carcinogens that result from digestion and excretion. The human large intestine has around 10^7 crypts that each renew about once per week. If a stem lineage in the human colon divided once every six days for 80 years, it

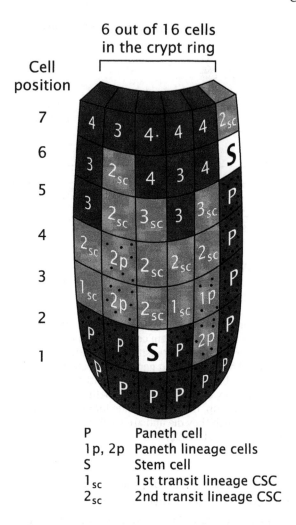

6 out of 16 cells in the crypt ring

Cell position

7	
6	
5	
4	
3	
2	
1	

P   Paneth cell
1p, 2p   Paneth lineage cells
S   Stem cell
1_{sc} 1st transit lineage CSC
2_{sc} 2nd transit lineage CSC

Figure 12.7 Three-dimensional schematic of a crypt in the mouse small intestine. The positions of the individual cells show how things might look in a typical crypt. The Paneth cells tend toward the bottom, where they contribute to innate immunity by responding to bacterial infection (Ayabe et al. 2000). The numbers on the cells show the transit cell generation i, as in the T_i of Figure 12.6. The stem cells vary in actual cellular position in the range 3–7, but on average appear to be around cell position 4 when numbered from the bottom. The figure only shows the bottom 7 cell positions of the approximately 15 positions. CSC abbreviates "clonogenic stem cell" (see Figure 12.6). Redrawn from Marshman et al. (2002).

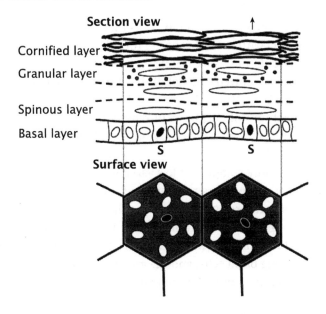

Figure 12.8   Architecture of skin renewal in the mouse based on Potten's model of epidermal proliferative units. The top cross-sectional view shows the epidermal layers. About one in ten basal cells are stem cells (S). The neighboring basal cells and all cells in the layers directly above derive from the stem cell in a typical stem-transit architecture. The surface view shows that each unit derived from a single stem cell forms a roughly hexagonal shape that encompasses about ten basal cells. Each black cell denotes the single stem cell in each unit. From Potten and Booth (2002).

would divide about 5,000 times. However, the actual history of stem lineages and the number of divisions over time remains unknown.

EPIDERMAL RENEWAL

The epidermal layer of the skin turns over about every 7 days in mice (Potten 1981; Ghazizadeh and Taichman 2001) and approximately every 60 days in humans (Hunter et al. 1995); however, those numbers must be taken only as rough estimates.

Several lines of indirect evidence suggest that the skin renews by a stem-transit architecture (Watt 1998; Janes et al. 2002). For example, about 60% of basal epidermal cells are progressing through the cell cycle, but in mice only about 10% of those cells can continue through several rounds of cell division after irradiation. Human epidermal cells plated in cell culture also show a distinction between rare cells that have a high capacity for self-renewal and common cells that divide only a few times. Those cycling cells with limited capacity for self-renewal are thought to be the transit population (Watt 1998).

Figure 12.8 shows Potten's model of the epidermal proliferation unit for mice (Potten 1974, 1981; Potten and Booth 2002). Each approximately hexagonal unit of surface skin renews from a basal layer comprising about ten cells, of which only one basal stem cell renews the unit.

Human skin is more complex: it has variable thickness in different locations, often has more layers than mouse skin, and has an undulating basal layer. Most authors agree that stem cells reside at the basal layer and give rise to an upward-migrating transit lineage. Controversy continues over the location of the stem cells in the basal layer, the frequency of stem cells among basal cells, and the architecture of stem-transit lineages and proliferative units (Potten and Booth 2002; Ghazizadeh and Taichman 2005).

The hairs in the epidermis renew by a different process. Figure 12.9 shows the hair cycle, in which each follicle alternates between rest and growth phases. During hair growth, there seems to be a stem-transit type of architecture: stem-like cells replace themselves in the follicular germ and simultaneously initiate transit lineages that move up and continue to divide. After the growth phase, the lower part of the follicle regresses.

It remains unclear where the stem cells come from to reseed the follicular germ at the start of the next growth phase. Those stem cells may come from cells in the follicular germ of the rest phase, shown as FG(s?) in Figure 12.9, or the next round of stem cells may migrate down from daughter cells produced by the stem cells in the bulge region. Potten and Booth (2002) emphasized the difficulty of interpreting various studies on this issue. Two recent studies favor the bulge stem cells as the progenitors for each new round of follicular growth (Morris et al. 2004; Kim et al. 2006).

In development, the stem cells of the bulge region appear to be the ultimate source for the interfollicular stem cells (those, for example, in Figure 12.8) and at least for the initial seeding of the follicular germ. After injury, the bulge stem cells can regenerate the hair follicle, sebaceous gland, and interfollicular proliferative units (Cotsarelis et al. 1990; Taylor et al. 2000; Potten and Booth 2002).

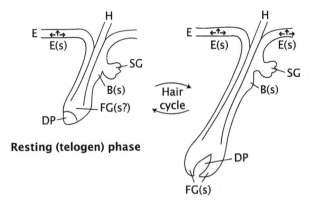

Resting (telogen) phase

Growing (anagen) phase

E	Epidermis	B	Bulge
H	Hair	SG	Sebaceous gland
DP	Dermal papilla	(s)	Stem cells
FG	Follicle germ		

Figure 12.9   Life cycle of a mammalian hair follicle. As the follicle moves from the rest phase to the growth phase, the follicular germ region moves downward and becomes an active site of cell division. Transit cells from the follicular germ move upward to form the growing hair. After a growth phase, the follicular germ region regresses to reform the rest phase morphology. From Potten and Booth (2002).

So far, I have discussed the keratinocyte lineages that produce the hair and the epidermal surface. In those tissues, melanocyte cell lineages provide pigmentation. Recent studies suggest that, in the hair follicles, the bulge region contains melanocyte stem cells (Nishimura et al. 2002; Lang et al. 2005; Sommer 2005). In each hair cycle, the melanocyte stem cells produce some daughters cells that migrate to the base of the follicle where the active keratinocyte transit lineages will be generated. Melanocytes in each new hair cycle seem to derive from the melanocyte stem cells in the bulge region.

Cancer risk concentrates in long-lived cell lineages—the stem lineages. Morris (2004) recently summarized evidence that various skin cancers derive from keratinocyte stem lineages. Similarly, melanomas probably descend from transformed melanocyte stem cells. Alternatively, transformed transit cells may de-differentiate into cancer cells with stem-like properties of renewal.

OTHER TISSUES

The blood, intestine, and skin renew frequently and have clear stem-transit architectures. Several other tissues also appear to have stem lineages that may provide a source for regular renewal, a reservoir for tissue repair, or daughter cell lineages that terminally differentiate (Lajtha 1979; Watt 1998).

Mammalian spermatogenesis has a clearly defined stem-transit architecture of renewal and differentiation (de Rooij 1998). In other tissues, the details of lineage history are less clear at present. Clarke et al. (2003) discuss a model of breast epithelium renewed by a stem-transit hierarchy of differentiation. Numerous recent articles describe the properties of breast stem cells (reviewed by Dontu et al. 2003; Liu et al. 2005; Villadsen 2005). Rizzo et al. (2005) discussed a stem-transit pathway of renewal for the normal prostate, but at present we have only limited understanding of tissue architecture in the prostate. Cells with some stem-like properties may occur in many tissues, but cell lineage architectures probably vary according to demands for cell turnover and regeneration.

12.3 Symmetric versus Asymmetric Stem Cell Divisions

To maintain a pool of N stem cells in a niche, each stem division must on average produce one daughter stem cell and one daughter that differentiates. Regulation of stem cell numbers may occur either by symmetric or asymmetric stem cell division (Cairns 1975; Watt and Hogan 2000; Morrison and Kimble 2006).

In symmetric division, each replication produces two identical daughter cells. Random processes then determine whether 0, 1, or 2 of the daughters remain stem cells while the other daughters differentiate. Over the whole pool of N stem cells, some process must regulate the probability of differentiation such that on average each stem division gives rise to one stem and one differentiated daughter.

In asymmetric division, the daughters differ. One daughter remains as a stem cell to replace the mother, and the other daughter differentiates.

The shape of cell lineages and the rate of evolutionary change in lineages depend on whether stem cells divide symmetrically or asymmetrically. I discuss those lineage consequences in the next section. Here, I briefly review evidence with regard to whether stem divisions are symmetric or asymmetric.

Several recent studies support the asymmetric pattern of stem cell division. Lechler and Fuchs (2005) showed in mice that dividing cells at the basal layer of the epidermis produce asymmetric daughters: one daughter moves upward while differentiating into a cell with limited proliferative capacity, whereas the other undifferentiated daughter remains at the basal layer and retains proliferative capacity. Asymmetric division of stem cells appears to split daughters between the stem and transit pathways. Asymmetry of daughter cell fate arises from asymmetry in the orientation of the mitotic spindles: one daughter moves upward from the basal membrane, and the other daughter remains near the basement membrane where it receives signals to maintain stem characteristics.

Drosophila spermatogenesis also divides its stem cells asymmetrically by mitotic spindle orientation and signals in the basal stem niche (Yamashita et al. 2003). It remains unclear whether mammalian sperm stem cells divide symmetrically or asymmetrically.

Preliminary in vitro evidence suggests that mammalian hematopoietic stem cells divide asymmetrically (Takano et al. 2004; Giebel et al. 2006); however, this hypothesis of hematopoietic stem cell asymmetry requires further analysis.

Although asymmetry seems to occur in a few particular cases, obtaining direct evidence of asymmetry remains technically challenging (e.g., Giebel et al. 2006). Another line of evidence in favor of asymmetry comes from the pattern by which DNA segregates to daughter cells.

12.4 Asymmetric Mitoses and the Stem Line Mutation Rate

Cairns (1975) emphasized that in a stem-transit architecture, only the stem lineage survives over time. Thus, only those mutations in the "immortal" stem lineage remain in the tissue. Cairns argued that organisms may use various mechanisms to reduce the mutation rate in the stem lineage.

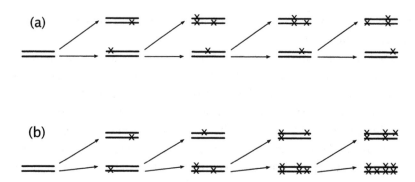

Figure 12.10 Cairns' (1975) hypothesis of asymmetric DNA segregation in stem cell divisions. (a) Immortal stranding, in which the stem lineage along the bottom always receives the older strand of the DNA duplex in each round of cell division. (b) Segregation of the newer DNA strand to the stem cell lineage in each round of cell division. Random segregation would follow a stochastic process between these two patterns. See text for full discussion.

IMMORTAL STRANDING

In every mitosis, the DNA duplex splits, each strand acting as a template for replication to produce a new complementary strand. It is possible that most mutations during replication arise on the newly synthesized strand. A stem lineage could reduce its mutation rate if each stem cell division segregated the oldest template strands to the daughter destined to remain in the stem lineage and the newer strands to the daughter destined for the short-lived transit lineage.

Figure 12.10a shows Cairns' hypothesis for segregation of DNA template strands. The DNA duplex at the lower left begins with identical DNA strands. The duplex splits as shown, and each strand serves as a template for replication. Suppose, each time a stem cell copies its DNA, that during replication one new mutation arises on the new strand. The "X" marks the new mutation. In the figure, the first round of replication shows the original templates without mutations and the newly replicated strands, each new strand with one mutation.

With each subsequent round of replication in Figure 12.10a, the older template without mutations segregates to the stem lineage along the bottom, and the younger strand with one new mutation segregates up to the transit lineage. This pattern reaches a steady state, in which the stem line retains the original template strand and a strand replicated once off the template with one new mutation. At the steady state, the transit lineage always receives a strand copied from the template that carries one new mutation; replication in the transit cell adds another mutation.

Figure 12.10b shows the opposite pattern, in which the newest strand always segregates to the stem lineage along the bottom. The newer strand always has one additional mutation, so the stem lineage accrues one new mutation in each generation.

By the standard view of DNA replication and mitosis, strands segregate randomly to daughter cells. If so, then the pattern by which mutations accumulate would follow a stochastic process between case (a), in which the stem lineage always gets the older strand, and (b), in which the stem lineage always gets the newer strand. Stochastic segregation would, on average, cause mutations to accumulate in the stem lineage at one-half the rate at which mutations arise on newly copied strands.

Cairns (1975) called the pattern in Figure 12.10a "immortal stranding." Any tendency away from purely random segregation and toward immortal stranding would lower the rate at which mutations accumulate in the stem line.

Immortal stranding requires asymmetric stem cell division, in which the fate of the daughters is determined during mitosis, before segregation occurs. Any evidence for immortal stranding also provides evidence for asymmetric stem cell division.

Several recent studies support Cairns' hypothesis of immortal stranding in stem cell lineages. Potten et al. (2002) marked DNA strands in mouse small intestine crypts with tritiated thymidine, then labeled newly synthesized strands with a different label, bromodeoxyuridine. Over time, only a few cells in crypt positions 3–7 retained the initial label; those cell positions delineate the crypt location in which stem cells reside (Figure 12.7). When the second label was removed, the putative stem cells that retained tritiated thymidine lost the second label, bromodeoxyuridine, showing that those cells did pass through the mitotic cycle.

Smith (2005) similarly showed that cells with stem lineage properties in mouse mammary glands retain immortal strands through epithelial tissue renewal.

Studies of asymmetrically dividing cells in tissue culture also demonstrate conditions under which immortal stranding occurs (Merok et al.

2002; Karpowicz et al. 2005). Interestingly, both asymmetric division and immortal stranding may be regulated by p53 and IMP dehydrogenase, the rate-determining enzyme in ribonucleotide biosynthesis (Rambhatla et al. 2005).

STEM CELL SENSITIVITY TO DNA DAMAGE

Mutations in the template strand of a stem cell carry forward through the stem lineage and the renewing tissue. Cairns (1975) suggested that if mutagens or other processes caused significant DNA damage to a stem cell, the cell might undergo apoptosis rather than risk repair. Apoptosis would reliably remove the mutations from the tissue. In particular, Cairns predicted that stem cells would be exceptionally prone to apoptosis in response to DNA damage when compared with other cells. Most other cells have a relatively short expected life for their descendant lineage; for those short-lived cell lineages, DNA damage does not impose such severe risks as for stem cell lineages.

Several studies suggest that stem cells have extreme sensitivity to damage, such that even a single radiation-induced hit can trigger apoptosis (Potten 1977; Hendry et al. 1982; Potten et al. 1992; Potten and Grant 1998). Those studies demonstrated sensitivity in gastrointestinal crypts near where stem cells reside, but it remains difficult to identify the exact location of stem cells in vivo.

We are left with an association between extreme radiosensitivity of a small fraction of cells and the expected location of stem cells. Potten et al. (2002) used the methods described above to label DNA strands and identify label-retaining cells as stem cells. They then found some evidence for an association between those cells that retain label and those cells that undergo apoptosis in response to mild radiation-induced damage.

TISSUE REPAIR AND RISK OF SYMMETRIC DIVISION

We can measure the age of a DNA strand as the number of strand replications back to some ancestral template. In Figure 12.10 each "X" on a strand measures age back to the ancestral template on the left.

If a stem cell dies, it may be replaced by another stem cell (Cairns 2002). The replacement requires a symmetric mitosis, because both daughters must be retained as stem cells in order to increase by one the number of stem cells in the pool. In a symmetric mitosis, the age of the DNA strands increases in one of the new daughter stem cells. This increase in age can be seen on the right side of Figure 12.10a. In a steady-state stem cell division, the top daughter that would normally segregate to the transit lineage has templates that have ages one and two relative to the initial template of age zero that the main stem lineage has retained.

The lost stem cell may alternatively be replaced by a daughter transit cell (Cairns 2002). If, for example, the most recent daughter transit cell on the right side of Figure 12.10a reverted to a stem cell, strand age would increase by one relative to the lost ancestral stem cell.

Mitogenesis caused by wounds, chemical carcinogens, or irritation increases the rate of cancer progression (reviewed by Peto 1977; Cairns 1998). Presumably wounds and other forms of tissue damage often kill stem cells; repair requires that those stem cells be replaced.

The interesting comparison is: How much of the increased risk comes from the accumulation of mutations in the stem line caused by symmetric mitoses, and how much of the enhanced risk comes from an increased rate of mitosis independently of the distinction between symmetry and asymmetry in DNA strand segregation?

12.5 Tissue Compartments
and Repression of Competition

The renewing epithelia of the intestine and skin have a compartmental structure (Figures 12.4 and 12.8). Each stem cell normally contributes only to its own compartment. This spatial restriction prevents competition between stem cell lineages in different compartments (Cairns 1975).

Suppose, for example, that a mutation caused a particular stem cell to replicate faster. That mutant lineage might take over its own compartment, outcompeting other stem lineages within the compartment. But spatial restrictions would often prevent the mutant lineage from spreading beyond its own small neighborhood. From a lineage perspective, each compartment limits the local population size and defines a separate parallel line of descent and evolution. An expanding clone, perhaps one step along in carcinogenesis, cannot normally expand beyond its compartmental boundaries, thus limiting the target number of cells for the accumulation of subsequent mutations.

Cairns (1975) pointed out that each tissue probably has different rules governing the territoriality of proliferating cells. Those spatial rules determine which kinds of variant cell succeed in each type of tissue. Those variants that could break territorial boundaries and invade neighboring compartments would gain a significant competitive advantage, increase their populations, and provide a large clonal target for subsequent advances in progression.

Repression of competition has become an important general concept in the study of cooperative evolution (Buss 1987; Frank 1995; Maynard Smith and Szathmary 1995; Frank 2003a). Perhaps such repression was an essential step in the evolution of complex multicellularity, in which large populations of independent cells act in a mostly cooperative manner.

12.6 Summary

This chapter reviewed the processes of tissue renewal. Most renewing tissues derive from a small number of stem cells. Mutations to stem cells pose the main risk for cancer. Stem cells may have various mechanisms to reduce their mutation rate. For example, the stem lineage may retain the DNA template and segregate new copies of the DNA to the daughter cells in the transit lineage. In addition, the patterns of tissue renewal from stem cells and the shape of stem cell lineages affect the accumulation of somatic mutations. To analyze in more detail how somatic mutations accumulate, I discuss in the next chapter the population genetics of somatic cell lineages.

13 Stem Cells: Population Genetics

Heritable changes in populations of cells drive cancer progression. In this chapter, I discuss three topics concerning population-level aspects of cellular genetics.

The first section shows that mutations during development may contribute significantly to cancer risk. In development, cell lineages expand exponentially to produce the cells that initially seed a tissue. A single mutation in an expanding population carries forward to many descendant cells. By contrast, once the tissue has developed, each new mutation usually remains confined to the localized area of the tissue that descends directly from the mutated cell. Because mutations during development carry forward to many more cells than mutations during renewal, a significant fraction of cancer risk may be determined in the short period of development early in life.

The second section analyzes the distinction between stem lineages and transit lineages. To renew a tissue, cells must be continuously produced to balance the equal number of cells that die. Cell death prunes certain cell lineages—the transit lineages—and requires that other lineages continue to provide future renewal—the stem lineages. Renewal imposes a constraint on the shape of stem and transit lineages. Within this constraint, if the mutation rate is relatively lower in stem cells, then relatively longer stem lineages and shorter transit lineages reduce cancer risk.

The third section contrasts symmetric and asymmetric mitoses in stem cells. Each stem cell may divide asymmetrically, every division giving rise to one daughter stem cell and one daughter transit cell. Alternatively, each stem cell may divide symmetrically, giving rise to two daughters that retain the potential to continue in the stem lineage; random selection among the pool of excess potential stem cells reduces the stem pool back to its constant size. With asymmetric division, any heritable change remains confined to the independent lineage in which it arose. With symmetric division, the random selection process causes each heritable change eventually to disappear or to become fixed in the stem pool; only one lineage survives over time.

13.1 Mutations during Development

Renewing tissues typically have two distinct phases in the history of their cellular lineages. Early in life, cellular lineages expand exponentially to form the tissue. For the remainder of life, stem cells renew the tissue by dividing to form a nearly linear cellular history. Figure 13.1 shows a schematic diagram of the exponential and linear phases of cellular division.

Mutations accumulate differently in the exponential and linear phases of cellular division (Frank and Nowak 2003). During the exponential phase of development, a mutation carries forward to many descendant cells. The initial stem cells derive from the exponential, developmental phase: one mutational event during development can cause many of the initial stem cells to carry and transmit that mutation. During the renewal phase, a mutation transmits only to the localized line of descent in that tissue compartment: one mutational event has limited consequences.

Development occurs over a relatively short fraction of the human lifespan. However, a significant fraction of cancer risk may arise from mutations during development, because the shape of cell lineage history differs during development from that in later periods of tissue renewal (Frank and Nowak 2003).

MUTATIONAL EVENTS VERSUS THE NUMBER OF MUTATED CELLS

Individuals begin life with one cell. At the end of development, a renewing tissue may have millions of stem cells. To go from one precursor cell to N initial stem cells requires at least $N - 1$ cell divisions, because each cell division increases the number of cells by one.

If the mutation rate per locus in each cellular generation is u, then how many of the initial N stem cells carry a mutation at a particular locus? This general kind of problem was first studied in microbial populations by Luria and Delbrück (Luria and Delbrück 1943; Zheng 1999, 2005). They wanted to estimate the mutation rate, u, in microbial populations by observing the fraction of the final N cells that carry a mutation.

The Luria-Delbrück problem plays a central role in the study of cancer, because progression depends on how heritable changes accumulate in cell lineages. The Luria-Delbrück analysis focuses on one aspect of

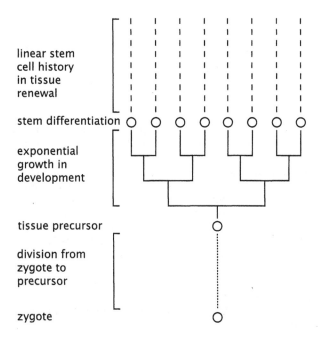

linear stem
cell history
in tissue
renewal

stem differentiation

exponential
growth in
development

tissue precursor

division from
zygote to
precursor

zygote

Figure 13.1   Lineage history of cells in renewing tissues. All cells trace their ancestry back to the zygote. Each tissue, or subset of tissue, derives from a precursor cell; n_p rounds of cell division separate the precursor cell from the zygote. From a precursor cell, n_e rounds of cell division lead to exponential clonal expansion until the descendants differentiate into the tissue-specific stem cells that seed the developing tissue. In a compartmental tissue, such as the intestine, lineage history of the renewing tissue follows an essentially linear path, in which each cellular history traces back through the same sequence of stem divisions (Figure 12.2). At any point in time, a cell traces its history back through n_s stem cell divisions to the ancestral stem cell in the tissue, and $n = n_p + n_e + n_s$ divisions back to the zygote. Modified from Frank and Nowak (2003).

mutation accumulation in cell lineages: the distribution of mutations in an exponentially expanding clone of cells.

To study the Luria-Delbrück problem, we must distinguish between mutational events and the number of cells that carry a mutation. Figure 13.2 shows an example in which one cell divides through three cellular generations to yield $N = 2^3 = 8$ descendants. This exponential growth requires a total of $N - 1 = 7$ cell divisions. Each cell division causes one cell to branch into two descendants, so there are $2(N - 1) = 14$ branches in which DNA is copied and a mutational event

8/14   4/14   2/14

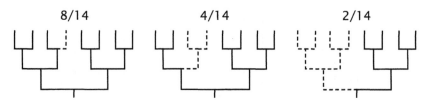

Figure 13.2   Probability of the number of mutated cells for a single mutational event. Each of the three sequences starts with a single cell that then proceeds through three generations of cell division, yielding $N = 2^3 = 8$ descendants. In each sequence, there are $2(N - 1) = 14$ branches, each branch representing an independent DNA copying process. I assume one mutational event with equal probability of occurring on any branch. On the left, there are 8 third-level branches, so the probability that the sequence yields one mutated cell is $2^3/[2(N - 1)] = 8/14$. In the middle, there are 4 second-level branches, so the probability that the sequence yields two mutated cells is $2^2/[2(N - 1)] = 4/14$. On the right, there are 2 first-level branches, so the probability that the sequence yields four mutated cells is $2^1/[2(N - 1)] = 2/14$. Early mutations in the sequence occur relatively rarely because there are fewer branches. When early mutations do occur, they carry forward to a large number of descendant cells; for this reason, the Luria-Delbrück distribution is sometimes called the jackpot distribution.

may occur. If one mutational event occurs among those 14 replications, then how many of the final 8 cells carry the mutation?

Figure 13.2 enumerates the possible outcomes for the simple example in which there is exactly one mutational event and a single cell divides regularly to produce 8 descendants. We can gain an intuitive understanding of the problem by generalizing the example in Figure 13.2.

Suppose we begin with one precursor cell, which then divides n times to yield $N = 2^n$ descendants. Assume that exactly one mutational event occurs, and that the mutational event happens with equal probability on any of the $2(N - 1)$ branches. If the mutation occurs on one branch in the first division, then $2^{-1} = 1/2$ of the descendants carry the mutation; if the mutation occurs on one branch in the second division, then $2^{-2} = 1/4$ of the descendants carry the mutation. In general, a fraction 2^{-i} of the descendants carries the mutation with probability $2^i/[2(N - 1)]$ for $i = 1, \ldots, n$ (Frank 2003b).

My simple calculations in the previous paragraph do not provide a full description of the Luria-Delbrück distribution, because I assumed exactly one mutational event over the entire population growth period. In reality, mutational events arise stochastically, so a full analysis must consider how the stochastic process of mutational events translates into the number of final cells that carry a mutation at a particular locus (Zheng 1999; Frank 2003b; Zheng 2005). For example, how do mutations during development translate into the number of initially mutated stem cells at the end of development?

Number of Initially Mutated Stem Cells

A small number of somatic mutations during development can lead to a significant fraction of stem cells carrying a mutation that predisposes to cancer. How much of the risk of cancer can be attributed to mutations that arise in development?

No one has tried to measure developmental risk. But a few simple calculations based on standard assumptions about cell division and mutation rate show that developmental risk may be important (Frank and Nowak 2003).

Suppose that $N = 10^8$ stem cells must be produced during development to seed the colon. Exponential growth of one cell into N cells requires about $\ln(N)$ cellular generations in the absence of cell death. In this case, $\ln(10^8) \approx 18$. If the mutation rate per locus per cell division during exponential growth is u_e, then the probability that any final stem cell carries a mutation at a particular locus, x, is roughly the mutation rate per cell division multiplied by the number of cell divisions, $x = u_e \ln(N)$. This probability is usually small: for example, if $u_e = 10^{-6}$, then x is of the order of 10^{-5}.

The frequency of initially mutated stem cells may be small, but the number may be significant. The average number of mutated cells at a particular locus is the number of cells, N, multiplied by the probability of mutation per cell, x. In this example, $Nx \approx 10^3$, or about one thousand.

I have focused on mutations at a single locus. Mutations at many different loci may predispose to cancer. Suppose mutations at L different loci can contribute to predisposition. We can get a rough idea of how multiple loci affect the process by simply adjusting the mutation rate per cell division to be a genome-wide rate of predisposing mutations, equal to $u_e L$. The number of loci that may affect predisposition may reasonably be around $L \approx 10^2$ and perhaps higher. Following the calculation in the previous paragraph, with $L \geq 10^2$, the number of initial stem cells carrying a predisposing mutation would on average be at

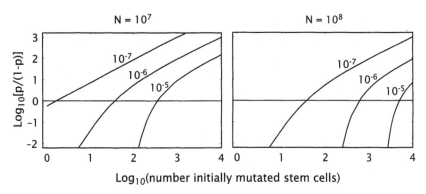

Figure 13.3 Number of initially mutated stem cells at the end of development. The total number of initial stem cells, N, derive by exponential growth from a single precursor cell. Each plot shows the cumulative probability, p, for the number of mutated initial stem cells. By plotting $\log_{10}[p/(1-p)]$, the zero line gives the median of the distribution. The number above each line is u_e, the mutation probability per cell added to the population during exponential growth. (I used an actual value of $10^{-5.2}$ rather than 10^{-5} because of computational limitations.) For a single gene, the mutation probability per gene per cell division, u_g, is probably greater than 10^{-7}. If there are at least $L = 100$ genes for which initial mutations can influence the progression to cancer, then $u_e = L \times u_g \geq 10^{-5}$. Initial mutations may, for example, occur in DNA repair genes, causing an elevated rate of mutation at other loci. Calculations made with algorithms in Zheng (2005). Modified from Frank and Nowak (2003).

least 10^5. Some individuals might have two predisposing mutations in a single initial stem cell.

The average number of initially mutated cells may be misleading, because the distribution for the number of mutants is highly skewed. A few rare individuals have a great excess; in those individuals, the mutation arises early in development, and most of the stem cells would carry the mutation. Those individuals would have the same risk as one who inherited the mutation.

Figure 13.3 shows the distribution for the number of initially mutated stem cells at the end of development. For example, in the right panel, with a mutation probability per cell division of 10^{-6}, a y value of 2 means that approximately 10^{-2}, or 1%, of the population has more than 10^4 initially mutated stem cells at a particular locus ($L = 1$). Similarly, with a mutation probability per cell division of 10^{-5}, a y value of 3 means that approximately 10^{-3}, or 0.1%, of the population has more than 10^4 initially mutated stem cells.

EXCESS RISK FROM DEVELOPMENTAL PREDISPOSITION

A significant fraction of adult-onset cancers may arise from mutations that occur during the short period of development early in life (Frank and Nowak 2003). In this section, I briefly summarize Meza et al.'s (2005) thorough quantitative analysis of this problem.

Meza et al. (2005) evaluated the role of developmental mutations in the context of colorectal cancer. They began with a model of progression and incidence that they had previously studied (Luebeck and Moolgavkar 2002). In that model, carcinogenesis progresses through four stages: two initial transitions, followed by a third transition that triggers clonal expansion, and then a final transition to the malignant stage.

In their new study, Meza et al. (2005) began with the same four-stage model. They then added a Luria-Delbrück process to obtain the probability distribution for the number of stem cells mutated at the end of development. The stochasticity in the Luria-Delbrück process causes a wide variation between individuals in the number of mutated stem cells. Meza et al. (2005) first calculated the probability that an individual carries Nx initially mutated stem cells at the end of development. To obtain overall population incidence, they summed the probability for each Nx multiplied by the incidence for individuals with Nx mutations.

Meza et al. (2005) summed incidence in their four-stage model over the number of initially mutated stem cells to fit the model's predicted incidence curve to the observed incidence of colorectal cancer in the USA. From their fitted model, they then estimated the proportion of cancers attributable to mutations that arise during development. Figure 13.4 shows that a high proportion of cancers may arise from mutations during the earliest stage of life.

Cancers at unusually young ages are often attributed to inheritance. However, Figure 13.4 suggests that early-onset cancers may often arise from developmental mutations. Developmental mutations act similarly to inherited mutations: if the developmental mutation happens in one of the first rounds of post-zygotic cell division, then many stem cells start life with the mutation. Inheritance is, in effect, a mutation that happened before the first zygotic division.

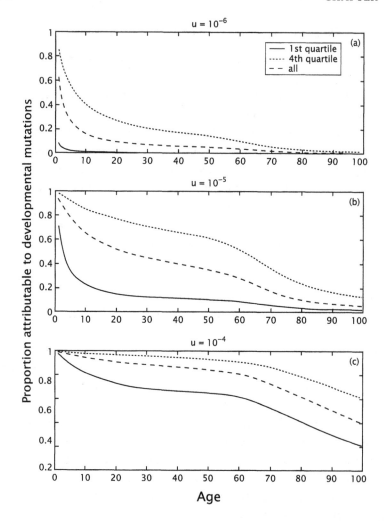

Figure 13.4   The proportion of cancers that arise from cells mutated during development. These plots show calculations based on a specific four-stage model of colorectal cancer progression (Meza et al. 2005). The parameters of the progression model were estimated from incidence data. The values of u above each plot show the mutation rate per year in stem cells. Stem cells likely divide between 10 and 100 times per year, thus a mutation rate per year of at least 10^{-5} per locus seems reasonable. In each plot, the three curves sketch the heterogeneity between individuals in risk attributable to developmental mutations. The first quartile shows the proportion of cancers at each age for those individuals whose risk is in the lowest 25% of the population, in particular, those individuals who by chance have the fewest stem cells mutated during development. Similarly, the fourth quartile shows the risk for the highest 25% of the population with regard to developmental mutations. From Meza et al. (2005).

Cell Generations to a Common Precursor Cell

When will cases with early onset and multiple tumors be caused by developmental mutations rather than inherited mutations? The answer depends on the pattern of cellular lineages that produce a tissue.

All cells in a tissue trace their ancestry back to a precursor cell. That common precursor would be the zygote if both cells from the first zygotic division contribute descendants to the tissue. Alternatively, several cell divisions derived from the zygote may occur before a precursor cell begins to differentiate into a particular tissue.

Figure 13.1 shows the different phases in the ancestry of a tissue. In that figure, n_p rounds of cell division happen between the zygote and the common precursor cell for the tissue. The precursor then seeds an exponentially growing clone through n_e cell generations. Once the tissue is formed, the stem cells renew the tissue by proceeding through n_s cell divisions, where n_s increases with age.

Consider an example to illustrate the potential importance of the number of cell generations to a common precursor for a tissue. Suppose a particular cancer syndrome has the characteristics of inherited disease—early onset and multiple independent tumors. Assume that the syndrome causes such severe early-onset disease that individuals who suffer the disease rarely reproduce. Then each case must arise from a new mutation.

The new mutation could occur in the parent: either in the germline or in a precursor to the germline that does not give rise to the affected tissue. A parental mutation would give rise to an inherited case, in which the offspring carries the mutation in all somatic cells. Suppose the number of cell generations between the parental germline precursor and the gamete is n_g. Alternatively, the new mutation could occur in the offspring. The number of cell generations between the zygote and the common precursor for the tissue is n_p.

The probability that an observed case arose from a developmental mutation rather than an inherited mutation would be approximately $n_p/(n_p + n_g)$. We could refine this approximation by adjusting for the mutation rates in the maternal and paternal germlines and the somatic precursor lineage and for the frequency of mutations carried by parents that derived from an earlier generation. For example, if the mutation rate per cell division is u, and the frequency of mutations carried by parents from earlier generations is f, then the approximation expands to $un_p/(un_p + un_g + f)$. My point here is simply that, as long as f is small, a significant fraction of important de novo mutations may happen developmentally rather than be inherited from parents.

Few estimates exist for n_p, the number of precursor cell generations. The little bit known about retinal development and the inherited cancer syndrome retinoblastoma raises some interesting issues. Retinoblastoma usually occurs before the age of five. Without modern medical treatment, the disease would often be fatal, so the affected individual would not reproduce. The inherited syndrome includes early onset and multiple independent tumors, usually with tumors in both eyes. According to the analysis here, the inherited syndrome would derive from developmental mutations approximately in a proportion of cases $n_p/(n_p + n_g)$.

The number of retinal precursor generations, n_p, remains unknown. Zaghloul et al. (2005) recently reviewed the subject of retinal development and concluded that, based on the *Xenopus* model, the left and right retina diverge rather late in development. Thus, there may be a significant number of precursor generations, n_p, before divergence of the common retinal precursors into the left and right eye. A developmental mutation before left-right divergence could predispose to bilateral retinoblastoma, a symptom usually attributed to an inherited mutation.

13.2 Stem-Transit Design

Mutations in transit cells usually get washed out as the transit cells slough at the surface (Cairns 1975). Most cell divisions occur in the transit lineages, and those divisions pose relatively little cancer risk. Because of the mutational washout advantage of transit lineages, it would seem that natural selection would favor a stem-transit separation with short-lived transit lineages. But adaptation may be more subtle.

Figure 13.5 shows the possibilities for design of a stem-transit architecture (Frank et al. 2003). Suppose a tissue requires k new cells over a certain period to renew itself. For now, assume that no other constraints exist with regard to renewal. To make k cells starting from one cell, the tissue may use n_1 stem cell divisions leading to n_1 transit lineages, each transit lineage dividing n_2 times to produce 2^{n_2} final cells, for a total of $k = n_1 2^{n_2}$ cells.

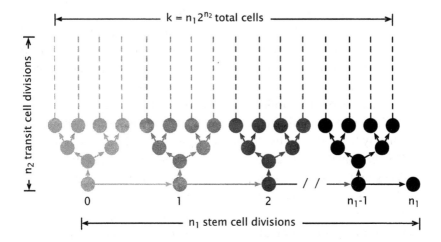

Figure 13.5   The pattern of cell division giving rise to a total of k cells. The single, initial cell divides to produce a stem cell and a transit lineage. Each transit lineage divides n_2 times, yielding 2^{n_2} cells. The stem lineage divides n_1 times, producing a total of $k = n_1 2^{n_2}$ cells. Redrawn from Frank et al. (2003).

Given the need to make k cells, consider how natural selection might increase benefit. Suppose short-lived transit lineages pose little risk. An improved design would add more cell divisions to those low-risk transit lineages and reduce the number of divisions in the long-lived stem lineage, that is, decrease n_1 and increase n_2.

In general, suppose we may choose to add one additional cell division to any lineage, with the goal to minimize cancer risk (Frank et al. 2003). If cancer requires n rate-limiting steps, and each step happens only during cell division, the risk rises with d^n, where d is the number of cell divisions. Risk increases exponentially with number of cell divisions in a lineage, thus natural selection favors prevention of long lineages. It is always most advantageous to add any new cell division to the shortest extant lineage. This optimal design maintains equal length among cell lineages.

In terms of tissue architecture, if we start with one cell, then the best design follows perfect binary cell division with all lineages remaining the same length, such that $k = 2^{n_2}$, where n_2 is the number of rounds of cell division. No stem divisions would occur except the first to seed the transit lineages.

This optimal design, with long transit lineages and no stem lineage, assumes that all k cells survive to the end of the required period, with no sloughing of cells. However, the requirement for continual cell death at epithelial surfaces imposes an additional requirement. But for now, I am just asking about the best design in the absence of the constraint imposed by renewal, to understand how much of tissue architecture may be explained by natural selection among alternative designs versus how much may be explained by the unavoidable constraints of renewal.

This first analysis suggests that natural selection favors long transit lineages and no stem lineage. If so, then the stem-transit design may be the consequence solely of continual cell death at the tissue surface, which imposes a stem-transit separation by shortening the cell lineages that lead to the sloughing of surface cells. But we should consider two additional factors.

First, the stem lineage may have a lower mutation rate than the transit lineage. Cairns (1975) proposed that immortal stranding and high sensitivity to DNA damage lower the stem-line mutation rate (See Section 12.4). If the stem lineage does have a lower mutation rate than the transit lineage, then natural selection would favor adding more cell divisions to the lower-risk stem line. In terms of design, this benefit of stem divisions would lengthen the stem lineage, that is, increase n_1 in Figure 13.5, and would shorten the higher-risk transit lineages, that is, decrease n_2.

Second, the transit lineage may be partially protected, because a transit cell that gets the required n carcinogenic changes may still slough off. This benefit would favor lengthening the transit lineages, because natural selection always tends to allocate additional divisions to those lineages with the lowest relative risk. This particular benefit for transit lineages works against the maintenance of a distinct, long-lived stem line.

In summary, two factors appear to favor a stem-transit design. A renewing tissue necessarily has continual cell death that prunes cell lineages and creates a dichotomy between short and long cell lineages. That constraint of tissue renewal may be sufficient to explain the stem-transit design, even though, with regard to cancer risk, natural selection often favors a more even distribution of cell lineage length. Alternatively, if the stem line accrues mutations at a lower rate than the transit

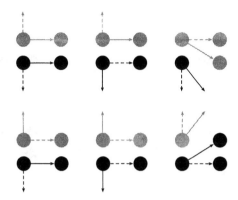

Figure 13.6   Stem-transit design to renew a tissue based on symmetric stem cell division and regulation of the stem pool to a constant size. Each alternative begins with two stem cells at the left. The two stem cells differ genetically. Each stem cell divides to produce two daughter cells; the solid versus dashed arrows represent the distinct daughter cells. The arrows up or down lead to transit cells; the arrows to the right lead to the replacement stem cells that remain at the base of tissue for subsequent renewal. There are six distinct patterns. The four patterns at the left retain genetic polymorphism in the stem pool and differ only in the four ways in which the distinct daughter cells can be assigned to stem or transit lineages. The two patterns at the right lose genetic variability in the stem pool; each of those patterns can happen in only one way, because the two daughters from each initial stem cell both move into the same compartment, either stem or transit, and so allow only one possible arrangement of daughter cells. Thus, with random choice of which cells remain in the stem pool, 4/6 of the time the polymorphism in the stem pool with be retained, 1/6 of the time the pool will become fixed for one genotype, and 1/6 of the time the pool will become fixed for the other genotype.

lines, then natural selection favors short transit lineages and long stem lineages.

13.3 Symmetric versus Asymmetric Mitoses

Suppose a tissue compartment, such as an intestinal crypt, maintains N stem cells. To maintain a constant stem pool size, each stem cell may divide asymmetrically, every division giving rise to one daughter stem cell and one daughter transit cell. Alternatively, each stem cell may divide symmetrically, giving rise to two daughters that retain the potential to continue in the stem lineage; random selection among the pool of excess potential stem cells reduces the stem pool back to N.

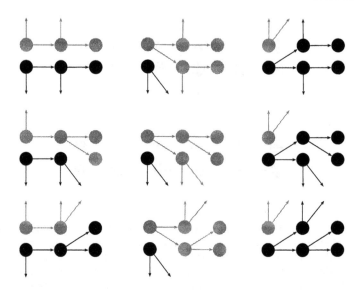

Figure 13.7   Symmetric stem cell division and regulation of the stem pool to a constant size by random selection of daughter cells. The three patterns in each generation—polymorphism, fixation for the light type, or fixation for the dark type—are shown in Figure 13.6. Here those three patterns are combined over two generations to form nine patterns. The probability for each pattern can be obtained by using the hypergeometric distribution. In general, if the stem pool size remains at N, and symmetric daughter cells migrate randomly to either the stem or transit pool, then starting with n black stem cells that double to $2n$, and m gray stem cells that double to $2m$, with $n + m = N$, the probability of retaining $0 \leq x \leq \alpha_n = \min(2n, N)$ black stem cells in the next pool of N is given by $P(x, n, N) = \binom{2n}{x}\binom{2m}{N-x}\big/\binom{2N}{N}$. Over two generations, $P_2(x, n, N) = \sum_i^{\alpha_n} P(x, i, N)P(i, n, N)$. From this formula, the probability of retaining polymorphism after two generations starting with $n = 1$ black cell and $N = 2$ stem cells is 16/36; the probability of ending with two black cells is 10/36; and the probability of ending with two white cells is 10/36.

With asymmetric division, the stem pool maintains N independent cell lineages. Any heritable change remains confined to the particular lineage in which it arose. The N distinct lineages form N parallel lines of evolution.

With symmetric division, the random selection process causes each heritable change eventually to disappear or to become fixed in the stem pool. In effect, only one lineage survives over many generation.

Figure 13.6 introduces a rough guide to the sorting of lineages under symmetric division. That figure shows a stem pool with $N = 2$, and the probability that the pool maintains two distinct lineages or coalesces into one lineage after a single round of cell division. Figure 13.7 calculates the probability of lineage diversity versus coalescence through two rounds of symmetric cell division.

Asymmetric and symmetric division have different consequences for the evolution of stem cell compartments. With asymmetric division, mutations remain in the stem pool but do not spread, unless those mutations break the asymmetry and force competition between lineages. With symmetric division, a mutation may be lost by chance or may take over the entire compartment. If a mutation takes over the compartment, any subsequent mutation in the compartment adds a second hit.

13.4 Summary

This chapter described the population genetics of somatic cell lineages, with an emphasis on stem cells. The theory of population genetics provides analytical tools to calculate how mutation, competition (selection), and random sorting of lineages (drift) influence the rate at which mutations accumulate in cell lineages. Several recent papers have applied population genetic theory to analyze how the demography of the stem cell compartment influences the accumulation of mutations and the progression of cancer (e.g., Komarova et al. 2003; Michor et al. 2003; Frank 2003c; Michor et al. 2004). The next chapter begins with empirical studies of stem cell population genetics, and follows with a more general review of cell lineage evolution and somatic mosaicism.

14 Cell Lineage History

The trillions of cells in a human slowly but steadily accumulate heritable change. Those heritable changes evolve in a spatially mosaic way. A few tissue patches may be advanced, poised to pass the next step to disease. Other tissue patches may be in an early stage, apparently normal but silently one step closer to malfunction.

Cancer progresses through heritable change to cells. Those heritable changes pass down cell lineages. To understand progression means to understand cell lineage history, and how different cell lineages interact.

New genetic technologies will soon provide vastly greater resolution in the measurement of heritable changes in cells: changes in DNA sequence, changes in DNA methylation, and changes in histone structure. Those new data will allow study of progression in terms of cell lineage history.

The first section discusses the reconstruction of cell lineage history from measurements of heritable changes in cells. The present studies remain crude, but hint at what will come. Variation in DNA methylation or repeated microsatellite sequences indicates the amount of heritable diversity among cells. Greater diversity suggests a longer time since the cells shared a common ancestor and a longer time in which the tissue has maintained independent cell lineages. By contrast, less diversity implies a shorter time to a common ancestor, perhaps caused by a recent clonal succession from a progenitor cell.

Measures of diversity suggest that colon crypts retain independent stem cell lineages for several years, but that clonal replacements occasionally homogenize the crypts. Crypts with *APC* mutations retain greater diversity, perhaps because those crypts retain independent lineages for relatively longer periods of time. Measures of diversity in hair follicles suggest that the follicles renew via a hierarchy of stem cells. The bulge region of the follicle contains ultimate stem cells that divide rarely, seeding the base of the hair with temporary stem cells that divide relatively frequently during each round of hair growth.

The second section analyzes how cell lineage history affects progression. Mitosis is known to be a key risk factor in cancer progression. The frequency of methylated DNA indicates the number of mitoses—a measure of mitotic age. I summarize data on the patterns of methylation with age in different tissues and discuss how those measures of mitotic age correspond to incidence patterns.

Another study tested the theory that progression develops through a series of clonal successions. Direct measurement of precancerous esophageal lesions found that progression to cancer increased with genetic diversity in the lesion. Greater genetic diversity may indicate a longer time since a common cellular ancestor and therefore less frequent clonal succession, contradicting the theory that clonal successions play a key role in progression.

The third section turns to measurements of somatic mosaicism, in which patches of cells carry an inherited change from a common ancestor. Mosaic patches may arise by a mutation during development or by a mutation in the adult that spreads by clonal expansion. Mosaic patches form a field with an increased risk of progression, in which multiple independent tumors may develop. At present, the best studies of mosaicism come from variants that cause visible skin defects, allowing direct observation of the altered tissue.

Genomic technologies can measure heritable changes in cells that lack an observable phenotype. Such genomic studies have already uncovered mosaicism in numerous tissues. Advancing technology will soon allow much more refined measures of genetic and epigenetic variation. Those measures will provide a window onto cell lineage history with regard to the accumulation of heritable change—the ultimate explanation of somatic evolution and progression to disease.

14.1 Reconstructing Cellular Phylogeny

Cell lineage histories affect the accumulation of heritable changes and the rate of carcinogenesis. For example, an expanding cell lineage poses significant risks, because a mutation carries forward to a growing clone of descendants. By contrast, the linear cellular history of renewal in an epithelial compartment poses lower risk, because a mutation carries forward only to the limited number of descendants in that single compartment.

One might think of this theory of cell lineages as a forward analysis: As time moves ahead, the pattern of descent influences the accumulation of heritable change and the progress of cancer.

In empirical studies, we often must consider the reverse: Given a set of cells that carry various heritable changes, how can we infer the ancestral lineage history of those cells? We know that, in an organism with a single-celled zygote, any two cells trace back to a common ancestral cell that is either the zygote or a descendant of the zygote. Similarly, any heritable change shared by a pair of cells often traces back to a common ancestor in which the original alteration occurred. Somatic changes trace back to a descendant of the zygote; inherited changes trace back to an ancestor of the zygote.

Evolutionary biologists have developed various methods to reconstruct the history of descent—the phylogeny (Page and Holmes 1998; Felsenstein 2003; Hall 2004). The methods essentially measure the relative likelihood of various ancestral relations between a set of cells, given the pattern of shared and variant characters in those cells. The characters may be DNA sequence, patterns of DNA methylation, or any other heritable characters.

An organism consists of a population of cells, whose cellular phylogeny describes its development and the lines of descent. Similarly, a tumor consists of numerous cells, in which the cellular phylogeny reflects the heritable changes that drove progression.

These points about cellular phylogeny have been known for a long time. But only recently has it been possible to reconstruct aspects of organismal history on the time scale of cellular generations.

I limit my discussion here to a few examples. I focus on cases that illustrate how phylogeny will help to understand the dynamics of progression and the patterns of age-specific incidence. This field will develop rapidly (Frumkin et al. 2005), but one can already outline some of the key concepts with regard to cancer dynamics and incidence (Shibata and Tavare 2006).

Variable Methylation Patterns

Epithelial cancers usually arise from the accumulation of heritable changes in stem cell lineages. The historical relations between the stem cells—their phylogeny—defines the shape of the cell lineage histories in which heritable changes accumulate. The phylogenetic shape influences the rate at which changes accumulate and therefore the dynamics of cancer progression.

The principles of lineage shape and progression are clear enough, but how can we figure out the actual history of stem cell lineages in an epithelial tissue compartment? In humans, one cannot use direct labeling or other invasive techniques, so to study the cell lineage histories, one must be able to read the past changes of cell lineages from the current differences between cells. From those current differences, one can infer how changes accumulated in the ancestral lineages that coalesce back to the common ancestor.

To infer phylogenetic history, one must study the right kind of character. If the character changes too slowly relative to the time scale of study, then the individual cells will not differ enough to infer historical relations. For example, if DNA point mutations happen at about 10^{-9} per base per cell division, then over a period of about 10^3 generations, one expects only one change per 10^6 bp in each cell relative to the common ancestor. With so little change, all of the extant cells would be nearly identical across sequences of up to 10^6 bp, and it would be impossible to reconstruct the history. At the other extreme, if characters change too fast, then the traces of history disappear.

In a normal gastrointestinal crypt, with up to 10^3 or so cell generations, standard DNA point mutations happen too rarely to reconstruct history with reasonable efficiency. To obtain sufficient information, Yatabe et al. (2001) measured methylation patterns. Adjacent DNA nucleotide sites that contain the bases C and G, linked by a phosphodiester bond and written CpG, may exist in a methylated or unmethylated state. Daughter cells inherit the methylation pattern of their parental cell. Random changes in the methylation state of each CpG pair happen roughly on the order of 10^{-5} per site per cell division (Shmookler Reis and Goldstein 1982; Pfeifer et al. 1990), which is much more frequent than point mutations in DNA sequence at about 10^{-9} per site per cell division.

NORMAL COLON CRYPTS

No one has yet reconstructed the phylogeny of cell lineages by study of methylation patterns. But Yatabe et al. (2001) developed a test of alternative shapes for stem cell lineages.

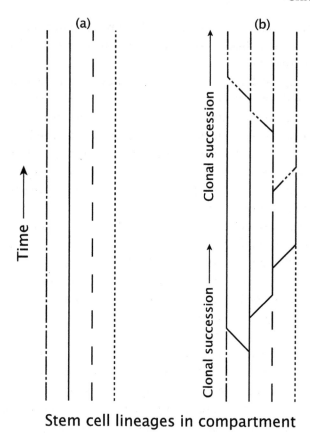

Stem cell lineages in compartment

Figure 14.1   Stem cell lineage history in a tissue compartment. (a) All stem cells division occur asymmetrically, maintaining each independent stem cell lineage. (b) Rare symmetric stem cell divisions lead to occasional loss of a stem cell lineage and replacement by another resident lineage. Over time, chance events cause loss of all lineages but one, leading to a sequence of clonal successions.

Yatabe et al. (2001) asked: Does a colon crypt maintain distinct stem cell lineages over time, or does a crypt proceed through a sequence of stem lineage replacements such that only one lineage survives over time? Figure 14.1 contrasts these alternatives. If stem cells always divide asymmetrically, then each stem cell division always produces one daughter stem cell to continue the lineage: the crypt maintains several distinct stem cell lineages. Alternatively, if occasionally a stem lineage failed to produce a daughter stem cell, that loss may be compensated by symmetric division of another stem lineage to produce two daughter stem cells.

Suppose purely asymmetric division occurs, and all independent stem lineages remain over time (Figure 14.1a). Then two lineages within a crypt will on average be as different as a pair of lineages sampled from different crypts. In each case, every lineage traces back to a different ancestral stem cell that seeded the colon crypts at the end of development.

Alternatively, suppose that occasional clonal succession occurs within crypts (Figure 14.1b). Then two lineages within a crypt will on average be more similar to each other than a pair of lineages sampled from different crypts. Within crypts, the current cells trace back to a recent common ancestor at the time of the last clonal succession. Between crypts, cells trace back to a more distant common ancestor that preceded the separation of the ancestral stem cells at the end of development.

Yatabe et al. (2001) showed that less variation in methylation occurs within crypts than between crypts, supporting the clonal succession model (Kim and Shibata 2002). Full evaluation of the data requires various assumptions about the number of stem cells per crypt, the rate of cell division, and the accuracy of the methylation measurement procedure (Ro and Rannala 2001). The overall conclusion of clonal succession appears to be well supported, but the estimated rate for clonal successions depends on several assumptions in the quantitative analysis. Based on those assumptions, Yatabe et al. (2001) infer that clonal succession happens on average about every 8.2 years.

COLON CRYPTS WITH AN INHERITED APC MUTATION

Inherited mutations to the *APC* locus cause familial adenomatous polyposis (FAP). In FAP, individuals may develop thousands of independently transformed crypts that lead to polyps or more aggressive tumors (Kinzler and Vogelstein 2002).

Mutations to *APC* play a role in stem cell dynamics (Kinzler and Vogelstein 2002). So Kim et al. (2004) hypothesized that those individuals who inherit an *APC* mutation may have altered patterns of stem lineage evolution in crypts when compared to normal individuals. To test this hypothesis, they compared the diversity of methylation patterns within crypts. Those crypts that carry germline *APC* mutations had higher methylation diversity than did crypts from normal individuals.

Methylation diversity in *APC*-mutated crypts may be higher because each stem lineage may survive longer, slowing down the rate of clonal succession. In Figure 14.1a, long-lived stem lineages retain methylation differences that arise in the separate lines, creating relatively high diversity over time. By contrast, in Figure 14.1b, each clonal succession drives out the diversity carried by the extinct lineages, keeping diversity low within the crypt.

Alternatively, *APC* mutations may increase methylation diversity by raising the number of stem lineages within a crypt. More stem lineages provide greater opportunity for the origin and maintenance of variation.

In either case, the greater methylation diversity in crypts with *APC* mutations signals that those crypts accumulate more genetic variation than normal crypts. Initially, that genetic variation may be neutral in the sense that it does not directly affect the survival or expansion of cell lineages. However, some of that variation may predispose to subsequent progression.

For example, a mutation to one allele of a tumor suppressor gene may have no consequences because the other, normal allele masks the effect of the mutation. But the hidden mutation poses a risk, because the next mutation to the normal allele knocks out function and may be a key step in progression (Nowak et al. 2002; Kim et al. 2004). So greater genetic diversity in crypts may itself be a predisposing risk.

STEM CELL HIERARCHY IN HAIR RENEWAL

Mammalian hair follicles renew throughout adult life. I described the hair renewal cycle in Section 12.2. Figure 14.2 reviews the main steps.

The cell lineage history within the hair follicle remains a puzzle (Potten and Booth 2002). One hypothesis suggests that, as a new hair cycle begins, stem cells in the bulge region divide, and their daughters move down to the follicular base to form the progenitors for the next round of growth. Those follicular progenitor cells act as the stem lineage during the growth phase, dividing to produce a transit lineage that moves up and forms the growing hair. As the growth cycle ends for that follicle, the follicular germ regresses to form the resting morphology (Figure 14.2).

If the follicular germ cells die off during regression, then the next round of growth must be seeded by new daughter cells from the stem

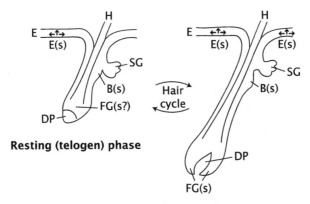

Resting (telogen) phase

Growing (anagen) phase

E	Epidermis	B	Bulge
H	Hair	SG	Sebaceous gland
DP	Dermal papilla	(s)	Stem cells
FG	Follicle germ		

Figure 14.2 Life cycle of a mammalian hair follicle. As the follicle moves from the rest phase to the growth phase, the follicular germ region moves downward and becomes an active site of cell division. Transit cells from the follicular germ move upward to form the growing hair. After a growth phase, the follicular germ region regresses to reform the rest phase morphology. From Potten and Booth (2002).

cells in the bulge region. That cycle would create a hierarchy of stem-transit lineages: bulge stem cells divide to start the cycle; daughters of the bulge cells form the follicular germ stem cells to feed the transit lineages for hair growth; the follicular germ stem cells die and the follicle regresses to resting morphology; the bulge cells divide again to start a new cycle. In this cycle, only the rarely dividing bulge lineage remains over time. Some evidence favors this stem cell hierarchy (Morris et al. 2004), but interpretation of the evidence remains ambiguous (Potten and Booth 2002).

Kim et al. (2006) analyzed methylation patterns of human hair follicles to evaluate the lineage history. Methylation patterns do not allow one directly to reconstruct the lineage history. Instead, one uses the fact that the frequency of methylated CpG nucleotide sites tends to increase with mitotic age—the number of cellular generations back to the zygote (Issa 2000; Yatabe et al. 2001). The actual methylation frequency in each cell varies stochastically but, on average, the methylation frequency provides an indicator of the number of cellular divisions.

If bulge cells divide rarely and continue to be the ultimate progenitors of hair renewal throughout life, then methylation will increase little with age. In particular, the average methylation of follicles should rise very early in life as cellular division during development creates the bulge stem cells, then follicular methylation should remain nearly constant during the remainder of life. Kim et al. (2006) found exactly that pattern: increasing methylation up to around two years of age, followed by a long plateau through the rest of life.

The bulge cells appear to be the ultimate stem cells in the follicle hierarchy. If so, then in each hair cycle the bulge cells seed the follicular germ with new daughter cells; those daughters act as stem cells for one cycle and then die.

During each cycle, the follicular germ cells divide, and their daughter transit lineages expand to produce the growing hair. The mitotic age of cells temporarily rises as the hair cycle progresses.

Kim et al. (2006) analyzed whether mitotic age measurably increases during a hair cycle by comparing methylation frequency between short and long hairs. Short hairs tend to be earlier in a given hair cycle than long hairs, and so the short hairs should on average have lower methylation frequency. The observed methylation patterns match this prediction of less methylation in short compared with long hairs. At the end of the hair cycle, the follicular germ apparently dies off, to be reseeded in the next cycle by relatively young and weakly methylated daughters of the bulge cells.

These particular conclusions about mitotic age and stem cell hierarchies remain tentative. The analysis does show clearly the potential value of inferring lineage history from molecular markers.

Variable Length of Microsatellite Repeats

Loss of DNA mismatch repair raises the mutation rate in repeated DNA sequences. One type of repeat, the microsatellite, mutates often in cells that are deficient in mismatch repair. I discuss two studies that measured variation in microsatellite repeats among a set of cells at one point in time, and used variation in those repeated regions to reconstruct historical aspects of the cell lineages involved in tumorigenesis.

Tsao et al. (1999) used microsatellite variability to reconstruct the cell lineage history of colorectal cancer progression in tissue that is deficient in mismatch repair. They tested two alternative scenarios for cell lineage history between tissues sampled from adenomas and adjacent cancerous outgrowths. Figure 14.3 shows the two alternatives.

In Figure 14.3a, the tissue progresses through repeated clonal successions. At any time, all cells derive from the common ancestor of the most recent clonal succession. Under this scenario, cells derived from the adenoma and the neighboring cancerous outgrowth have a relatively recent common ancestor.

In Figure 14.3b, clonal successions occur rarely. Instead, the tissue retains multiple distinct lineages. Under this scenario, cells derived from the adenoma and the neighboring cancerous outgrowth have a relatively distant common ancestor.

Tsao et al. (1999) tested these alternatives by measuring relative times as follows. Loss of mismatch repair (MMR$^-$) initiates an increased mutation rate that speeds cancer progression and also increases the rate of microsatellite mutations. By comparing the microsatellites of the adenoma and cancer samples with other tissues, one can estimate the total accumulation of microsatellite variation since the loss of mismatch repair. The microsatellite variation between the adenoma and cancer samples can then be scaled relative to the total variation, providing an estimate for the relative timing of the adenoma-cancer split compared with the loss of mismatch repair.

Figure 14.4 shows samples from two patients. The lineage history of each patient closely matches the hypothetical pattern of Figure 14.3b, in which the lineages derive from a relatively distant ancestor. Those observations support the hypothesis that colorectal cancer progression can retain several independent lines of progression following a key initiating event, in this case, loss of mismatch repair.

14.2 Demography of Progression

Changes in the age-onset curve of cancer measure the causal effect of carcinogenic processes. In this regard, different cell lineage histories have significant consequences to the extent that they alter age-specific incidence. I discuss a few examples.

(a) The common ancestor shifts with clonal succession

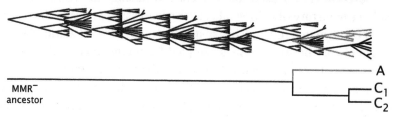

(b) Early common ancestor preserved with multiple lineages

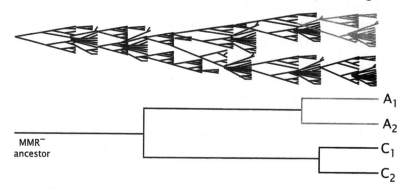

Figure 14.3 Alternative hypotheses for the relative time to the common ancestor between cells from a colorectal adenoma and an adjacent cancer. Time is measured relative to the ancestral loss of a component of DNA mismatch repair (MMR$^-$). Lightest gray shows the ancestry of cells sampled from the current adenoma. Black shows the ancestry of cells sampled from the current cancer. (a) Successive clonal expansions during multistage progression continually move the most recent common ancestor within the tissue to the recent past. Thus, the divergence is recent between the samples of the remaining adenoma tissue (A) and the developing cancer (C$_1$ and C$_2$) when compared with the time back to the ancestral loss of MMR. (b) After the initial MMR$^-$ event, the tissue retains multiple independent lines of progression. At diagnosis, two samples from the remaining adenoma (A$_1$ and A$_2$) derive from a relatively recent ancestor, and similarly, two samples from the developing cancer (C$_1$ and C$_2$) also derive from a recent ancestor. By contrast, cells from the adenoma and cancer derive from a more distant common ancestor, relatively close to the original MMR$^-$ mutation. Redrawn from Tsao et al. (1999).

MITOTIC AGE

Mitosis is perhaps the greatest risk factor in carcinogenesis (Peto 1977; Preston-Martin et al. 1990; Ames and Gold 1990; Cairns 1998).

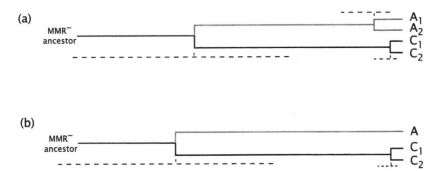

Figure 14.4   Reconstruction of cell lineage histories from samples of adenoma and cancer tissues in two patients. The accumulation of microsatellite variation caused by mismatch repair deficiency (MMR⁻) measures time in proportion to the number of cell generations. The lengths of the branches represent inferred time. The dashed lines show the estimated 95% confidence intervals for the timing of the branch points. (a) Samples from an adenoma and an adjacent cancerous outgrowth. The branch between the adenoma and the cancer happens fairly far back in the cell lineage history, as in Figure 14.3b, supporting the pattern of multilineage progression following MMR loss rather than frequent clonal successions. (b) The adjacent adenoma and cancer samples again suggest a fairly distant common ancestor, supporting multilineage progression since the origin of the MMR⁻ phenotype. Redrawn from Tsao et al. (1999).

Cell division induces new heritable variants and provides the opportunity for cellular competition and selection. The number of cell divisions in a lineage—mitotic age—provides a simple summary statistic of lineage history (Shibata and Tavare 2006).

Age-specific rates of mitosis can influence the age of cancer onset. In tissues such as the retina or the bones, mitosis and cancer happen relatively frequently early in life but rarely in adults. By contrast, renewing epithelial tissues in the colon and lung suffer increasing rates of cancer as the number of mitoses rises with age.

METHYLATION MEASURES MITOTIC AGE

Retina and bone differ qualitatively in mitotic pattern from colon and lung. These contrasting tissues lead to obvious comparisons in incidence. In other tissues, it may be difficult to guess the age-specific patterns of mitosis. So, Shibata and colleagues took the next step, by empirically estimating the number of lifetime mitoses in a lineage—mitotic age—from DNA methylation patterns.

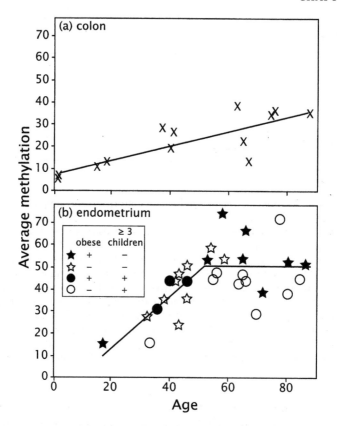

Figure 14.5 DNA methylation measures mitotic age. (a) Methylation increases steadily in the colon with chronological age, reflecting the continual mitosis in this tissue throughout life. (b) In the endometrium, methylation increases steadily with age during menstrual cycling and then plateaus after menopause. Obese women (filled symbols) have higher estrogen levels and greater endometrial turnover than non-obese women (open symbols). Obese women also have greater methylation than non-obese women, supporting the idea that methylation measures number of mitoses: 7 of 8 obese samples fall above the line, whereas 11 of 17 non-obese women fall below the line. Women with fewer than three children (stars) have more menstrual cycles and endometrial renewal than women with three or more children (circles). Women with few children have greater methylation: 11 of 14 women with less than three children fall above the line, whereas 9 of 11 women with more than three children fall below the line. Redrawn from Shibata and Tavare (2006), based on original studies in Yatabe et al. (2001) and Kim et al. (2005).

Various lines of evidence show that the frequency of methylation at certain genomic sites increases with the number of mitoses (Shibata and

Tavare 2006). For example, Kim et al. (2006) measured mitotic age in hair follicles by the frequency of CpG methylation, as I discussed in an earlier section.

Two studies report the age-specific frequency of methylation in the colon and endometrium (Yatabe et al. 2001; Kim et al. 2005). The colon shows a continuous rise in methylation frequency with age (Figure 14.5a). That steady rise in methylation supports the usual view of the colon as a continuously renewing tissue throughout life.

By contrast, methylation of the endometrium increases sharply to the age of menopause, then levels off through the remainder of life (Figure 14.5b). The early-life rise in endometrial methylation corresponds to the period of menstrual cycling and frequent tissue renewal. The late-life plateau corresponds to the period of reproductive quiescence and limited turnover of reproductive tissues.

Two further observations support the hypothesis that methylation correlates with mitotic age. Obese women typically have higher estrogen levels and greater reproductive tissue renewal than do lean women; obese women had correspondingly higher methylation levels of endometrial tissue than lean women (Figure 14.5b). Women with two or fewer children typically have more lifetime menstrual cycles than do women with three or more children; those women with fewer children and more menstrual cycles had correspondingly higher levels of methylation than those women with more children (Figure 14.5b).

MITOTIC AGE AND INCIDENCE IN FEMALE REPRODUCTIVE TISSUES

If mitoses do in fact drive progression, then the patterns of age-specific mitosis should correspond to patterns of age-specific incidence. Pike et al. (1983) argued that reduced mitotic rate of the breast after menopause causes the observed drop in the slope of the age-specific incidence of breast cancer later in life. Pike et al. (2004) updated the analysis to include the slowing rate of increase in cancer of the breast, ovary, and endometrium later in life. In Pike's formulation of the theory, incidence increases with mitotic age, so the rise in incidence for female reproductive tissues slows later in life as mitosis slows.

I use log-log acceleration (LLA) to measure the change in incidence with age. Figure A.8 shows plots of LLA for ovarian cancer. Those LLA curves follow the declining acceleration through the later part of life described by Pike. Figure A.2 shows that breast cancer also has declining acceleration with age. Those declining LLA curves fit Pike's prediction. However, notice in Figure A.8 that cancers of the kidney, esophagus, and larynx have declining patterns of LLA that closely match the pattern of decline for ovarian cancer.

The slowing of mitosis with age in the female reproductive tissues may very well reduce the LLA of those tissues. But the fact that non-reproductive tissues show similar declines suggests that ubiquitous aspects of aging may dominate the patterns of incidence.

MATHEMATICAL ANALYSIS OF MITOTIC AGE AND INCIDENCE

Pike developed a mathematical expression to link mitotic age to incidence (Pike et al. 1983, 2004). That formulation arises from the correct notion that the age-specific rate of mitosis may influence age-specific incidence. However, Pike's particular formulation incorrectly expresses the relation between mitosis and incidence. In this section, I show Pike's formulation, explain why it is wrong, and discuss the correct way to analyze the problem.

I begin by following Pike's formulation, but I modify the notation to match mine. Pike began with the widely used approximation for incidence

$$I(t) \approx ct^{n-1},$$

where t is time since birth, or, equivalently, age, and c is a constant that absorbs all terms independent of age. This formulation assumes that the risk factors driving cancer happen at the same constant rate throughout life. If mitosis is the main risk factor, and the rate of mitosis varies with age, then instead of measuring the accumulation of time by t, one should measure the accumulation of mitoses over time, or mitotic age, $m(t)$, where m is the cumulative number of mitoses at age t.

Pike therefore substituted mitotic age for age and presented the formula

$$I(t) \approx c[m(t)]^{n-1}. \tag{14.1}$$

This formulation is incorrect. For example, suppose that the age-specific rate of mitosis slows to near zero at age 65. The cumulative number of mitoses since birth at age 65, $m(65)$, may be a large number, and so according to Pike, the incidence will be high at age 65. However, incidence at age 65 is the rate of new cases at that age. If mitoses have slowed to almost zero, then this particular form of multistage theory predicts that new cases will be very rare, and incidence at age 65 should be near zero.

Let us suppose, for the moment, that it is possible to use mitotic age to obtain an approximation for incidence along the lines followed by Pike. How would we proceed to get the correct formulation? Start by assuming that the rate of mitosis determines the rate of transition between stages in a multistage model. Let the rate of mitosis at age t be $u(t)$. Then mitotic age at age t is the cumulative number of mitoses at that age, $m(t) = \int_0^t u(x)dx$, where the integral simply means the summing up of all the mitoses between ages 0 and t.

This measure of mitotic age describes the cumulative number of mitoses, so we need to work with cumulative incidence to keep cause and effect on the same scale. Cumulative incidence is the summing up of incidence between ages 0 and t, which is notationally $CI(t) = \int_0^t I(x)dx$. Then the widely used approximation that Pike wished to analyze is

$$CI(t) \approx c\,[m(t)]^n.$$

Age-specific incidence, $I(t)$, is the rate of additional cases at age t, which is the derivative of cumulative incidence with respect to t. Taking the derivative of both sides of the expression for cumulative incidence yields

$$I(t) \approx c\,[m(t)]^{n-1}\,u(t), \qquad\qquad (14.2)$$

which is the correct formula that follows from Pike's logic instead of Eq. (14.1). This correct formula can be read as: the rate of cancer at age t depends on mitotic age, $m(t)$, raised to the $n-1$st power, multiplied by the age-specific rate of mitosis at age t, $u(t)$. Mitotic age raised to the $n-1$st power is approximately proportional to the number of individuals that have progressed through the first $n-1$ stages of carcinogenesis and need only one additional step to be transformed into a case of cancer. The rate of mitosis at age t, $u(t)$, is the rate at which those individuals in stage $n-1$ pass the final step and are transformed. If the age-specific rate of mitosis, $u(t)$, drops significantly at menopause, then the incidence would also decline significantly, and the slope of the incidence curve would be negative. Pierce and Vaeth (2003) developed this sort of formulation properly and extensively for a period of carcinogen exposure followed by cessation of exposure. In that formulation, incidence depends on cumulative exposure and the current exposure rate, instead of cumulative mitoses and the current mitotic rate.

Although Eq. (14.2) is the right idea, the approximation will in fact often be highly inaccurate. The actual incidence at each age depends on the distribution of individuals in particular stages of progression. When rates of transition between stages change with age, the distribution of individuals in particular stages becomes particularly distorted with regard to the approximation in Eq. (14.2), which assumes a regular distribution pattern. For these reasons, I always advocate a direct calculation of the exact pattern of incidence, which can easily be accomplished for almost any set of assumptions, as explained in the earlier theory chapters. I took the trouble here to step through the difficulties encountered by Pike's analysis, because his approach and the associated problems occur often in the literature.

CLONAL SUCCESSION VERSUS MULTILINEAGE PROGRESSION

Frequent clonal expansion during progression causes cells to share a recent common ancestor. By contrast, less frequent clonal expansion allows different lineages to persist and differentiate over time. Do the early stages of carcinogenesis proceed by successive rounds of clonal expansion or by persistence of multiple lineages?

I discussed two relevant studies earlier in this chapter. Kim et al. (2004) showed a correlation between multilineage persistence and cancer risk. In their study, inherited *APC* mutations caused colon crypts to maintain more genetic diversity than crypts without such mutations. Kim et al. (2004) interpreted the greater diversity to mean that different cells in *APC*-mutated crypts traced their ancestry back to a more distant common ancestor than did cells in crypts that lack *APC* mutations. Greater multilineage persistence and genetic diversity in *APC*-mutated crypts correlate with a higher rate of cancer, but no evidence directly links diversity and lineage persistence to progression.

Tsao et al. (1999) studied cell lineage history between tissues sampled from colorectal adenomas and adjacent cancerous outgrowths. They analyzed cases in which the tissues had lost DNA mismatch repair, a key initiating event in carcinogenesis. They found that two patients apparently maintained distinct cell lineages during much of the time course of progression (Figure 14.4). Those observation are consistent with multilineage progression rather than frequent clonal succession.

Maley et al. (2006) provide further support for the association between genetic diversity in precancerous lesions and progression. They studied Barrett's esophagus, a premalignant lesion that often covers several centimeters of tissue and is too large for complete removal. Multiple biopsies provided several tissue samples per individual.

Maley et al. (2006) measured a variety of morphological and genetic attributes from each patient. Greater size of the premalignant lesion provided a weak but significant predictor for the risk of progression to malignancy. Indicators of genetic instability—loss of heterozygosity at *p53* and ploidy abnormalities—provided strong predictors for the risk of progression.

Genetic diversity within a lesion also provided a strong predictor of progression. At least two different hypotheses may explain why some lesions have more genetic diversity than others. First, mutations happen more often in some lesions than in others. Second, lineages may trace back to more distant ancestors in some lesions than others, allowing more time for diversity to accumulate.

To test whether progression depended only on mutation rate rather than lineage depth, Maley et al. (2006) calculated the effect of genetic diversity while controlling for indicators of genetic instability and mutation rate. They found that genetic diversity had a strong effect independently of indicators of mutation rate, suggesting that diversity caused by deep lineages correlates with progression.

From these observations on Barrett's esophagus, Maley et al. (2006) and Shibata (2006) conclude that the maintenance of multiple independent lineages accelerates progression. It may be that each clonal succession drives out genetic variability in a tissue, reducing the opportunity for future mutations to create combinations of genes that promote carcinogenesis.

All of these examples provide only indirect support for multilineage progression; they certainly do not rule out the importance of clonal expansion in progression. But remember that we are just in the very first years during which technology allows direct measurement of genetic variation in samples of tissues. Advances in technology will eventually provide better reconstructions of cell lineage history (e.g., Backvall et al.

2005). Such reconstructions will open a new window onto the dynamics of progression.

CLONAL EXPANSIONS AND CANCER STEM CELLS

Clonal expansion gives rise to a population of cells. Those cells may be in a precancerous state, ready to make the next transition along the pathway of progression. Or those cells may form a malignant tumor that will continue to grow and evolve.

In a clonal population, what fraction of the cells retain the potential to be the progenitors of future cell lineages? Put another way, what fraction can act as the stem cells that renew the population?

Some studies suggest that only a small fraction of cells in a tumor retain the potential to renew the population—the cancer stem cells (Reya et al. 2001; Pardal et al. 2003; Huntly and Gilliland 2005; Bapat 2006). Little information exists about earlier stages in progression.

Suppose, in an early precancerous clonal expansion, only a small fraction of the cells can act as long-term progenitors. Then, in spite of the large population of cells in the clone, only a small number of cells may drive progression to the next stage along the pathway to cancer. So clonal expansions do not necessarily raise the target size for future transitions and the rate of progression. What matters is the number of cells that retain the potential to be long-term progenitors.

14.3 Somatic Mosaicism

In each cell division, new heritable changes may arise in DNA sequence, in DNA methylation, and in modifications to histone proteins. A change in the first few post-zygotic divisions alters many descendants; a change in an epithelial stem cell modifies the descendants within the local tissue compartment. In either case, the organism develops into a mosaic of different genotypes.

Most observations of mosaicism derive from some spectacularly noticeable change. Pigmented skin patches mark the bounds of mosaic regions. A tumor emerges from several heritable changes in a region. Sometimes, multiple tumors develop within a broader field of altered cells.

Pigmented skin patches and tumors are rare, but mosaicism may be common. As individuals age, different tissue regions progress through

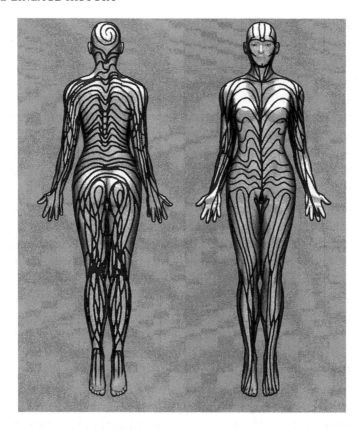

Figure 14.6   Epidermal skin aberrations often follow the lines of Blaschko. This pattern frequently traces mosaic cells that carry heritable aberrations, but the particular genetic or epigenetic modifications have not been described for all diseases (Taibjee et al. 2004; Chuong et al. 2006; Siegel and Sybert 2006). Drawing by Davide Brunelli (http://www.med- ars.it), reprinted with permission.

early, invisible stages of carcinogenesis. As genetic technologies improve, we will be able to measure the hidden mosaic evolution of cell lineages that drives cancer progression.

In this section, I mention some readily apparent cases of mosaicism. Those examples hint at the hidden processes of progression and at what we may learn in the near future.

Developmental Mosaicism

A single mutational event during development transmits through descendant cell lineages to create a large mosaic population. Each cell division during development may, on average, suffer a high probability of creating at least one heritable change. Among the trillions of cells in a human, each of billions of different heritable changes forms its own distinct mosaic pattern.

Gottlieb et al. (2001) list 30 diseases with reported mosaicism. I briefly discuss skin disorders with visible phenotypes, because the altered skin markings provide the easiest examples for study (Happle 1993).

The spatial distribution of the mosaic cells traces the tips of the cell lineage trees. If descendant cells remain together, then the mosaics form patches of distinct cells. In some cases, the descendant cell lineages trace distinctive patterns that reflect the movement of cells during development. For example, several visible skin diseases follow the lines of Blaschko (Figure 14.6). Other distinct patterns also occur in skin diseases (Chuong et al. 2006). Speckled lentiginous naevus and Becker's naevus follow a mosaic checkerboard pattern; mosaic trisomy of chromosome 13 causes scattered leaf-like shapes of hypopigmentation.

Familial glomuvenous malformations provide an excellent system to study the process of developmental mosaicism. These venous malformations appear on the skin as blue-red nodules (Vikkula et al. 2001; Brouillard and Vikkula 2003).

Individuals who inherit a mutation to one of the two alleles at the *glomulin* locus develop multiple independent nodules distributed randomly across their skin. By contrast, noninherited cases typically arise as a single, isolated nodule (Rudolph 1993; Boon et al. 1999; Happle and Konig 1999; Brouillard et al. 2002).

These patterns suggest that nodules form when both alleles of the *glomulin* locus have lost function. In inherited cases, the spatial pattern of nodules likely marks the multiple independent inactivations of the second allele at different locations during development (Happle and Konig 1999; Happle 1999; Brouillard et al. 2005). Study of individuals who inherit one nonfunctional allele would provide interesting data on developmental mosaicism. The number, spatial distribution, and size of nodules would describe the loss of the second allele, either by direct mutation, loss of heterozygosity, or epigenetic silencing.

The nodules of glomuvenous malformations record mutational events in the cell lineage history of the developing organism. Those events focus on a single locus. Independent heritable changes also accumulate at thousands of other genes. Of those thousands of genes, several hundred affect DNA repair and chromosomal maintenance; probably several hundred other loci control the cell cycle and cell death.

Heritable changes in any of those hundreds of DNA repair or cell-cycle genes may advance cancer progression through the first stages. Simple calculations suggest that such developmental mosaicism may contribute significantly to the incidence of cancer (Frank and Nowak 2003; Meza et al. 2005).

Fields Derived from Clonal Expansions

Mutations during development can create a population of descendant cells that have progressed toward cancer. Alternatively, in a fully developed individual, a single mutated cell may expand clonally to create a local patch or field of tissue that has progressed through an early stage of carcinogenesis. Most reported cases of a precancerous field do not distinguish between developmental mutations and clonal expansions in the fully developed organism.

Slaughter et al. (1953) introduced the idea that localized tumors may emerge from a broader precancerous field. Subsequent work on "field cancerization" almost always assumes that the field grows by clonal expansion of a mutated cell in the fully developed organism (Braakhuis et al. 2003, 2005; Hunter et al. 2005).

Several different lines of evidence may indicate a broader field surrounding a localized tumor: neighboring tissue may be histologically abnormal; genetic analysis may directly measure the spatial distribution of a mutated gene; and multiple independent tumors may develop from the same tissue patch. Improved genomic technologies make it increasingly easy to use direct genetic analysis. Those genetic analyses often demonstrate a broad field containing the same clonally derived mutation in tissue that appears normal.

Fields of *p53* mutants have been observed in the bladder (Simon et al. 2001), oral cavity (Braakhuis et al. 2003), and skin (Jonason et al. 1996; Brash 2006). Fields have also been observed in the lung, esophagus, vulva, cervix, colon, and breast (reviewed by Braakhuis et al. 2003). The importance of fields in progression depends on the fraction of cells in the expanded clone that retain the ability to progress further. It may be that only a limited fraction of cells retain or could acquire the stem-like properties needed for progression.

SPATIOTEMPORAL VARIATION IN PROGRESSION

Several developmental mutant patches appear in the case of inherited glomuvenous malformations; *p53* mutant patches can often be detected in normal tissue. Those fields form large, easily studied mutant patches on readily accessible surface tissues. Many more mutants must exist throughout normal tissue, in the hundreds of other genes that can affect progression.

In other words, the organism evolves continually in a mosaic way. Patches of varying size progress to different stages on the pathway to disease. Current data on evolving mosaicism focus on a few genes in a few tissues, measured over broad tissue patches. Soon, technology will allow measurement of more genes at finer spatial scales. With such data, we will begin to infer cell lineage histories with regard to the accumulation of heritable change. The cell lineage histories provide the ultimate explanation of somatic evolution and progression to disease. The diseases affected by somatic evolution may go beyond cancer, to include various syndromes that increase with age (Wallace 2005).

14.4 Summary

This chapter reviewed recent studies on the somatic evolution of cell lineages. Because cancer arises from the accumulation of heritable changes in cell lineages, such studies will play a key role in future analyses of cancer progression. Advancing genomic technologies will soon yield much greater resolution in the measurement of heritable cellular changes. To interpret those data, we will have to understand how such changes influence the dynamics of progression and the patterns of age-specific incidence. Shifts in incidence curves provide the ultimate measure of causation in cancer.

15 Conclusions

Molecular technology promises to reveal the biochemical changes of cancer. With that promise has also come an implicit assumption: one will understand cancer by enumerating the major biochemical changes involved in progression and the linkages of biochemical processes into networks that control cellular birth and death. But enumerating parts and their connections is not enough.

Think about a large airplane. If you were on that plane, the flight trajectory is what you would most care about. Could you predict the flight trajectory if you knew all of the individual control systems and their complex feedbacks? Probably not, because an inventory by itself does not provide all of the rates at which changes occur. Even with all of the rates for component processes, it would not be easy to work out the trajectory.

One needs to link the parts to the outcome: how do particular changes in components shift the plane's trajectory? One ultimately assigns causality to parts by how changes in the parts affect changes in the outcome.

In a similar way, a genetic or environmental factor causes cancer to the extent that it shifts the age-incidence curve—the trajectory of cancer. To understand a particular type of cancer, we must understand the forces that shape the age-incidence curve and the forces that shift the curve from its normal pattern.

This book developed a synthesis between, on the one hand, the biochemical processes that control cells and tissues, and, on the other hand, the consequences for the age-incidence curve of cancer. There have, of course, been many attempts to connect biochemistry to progression dynamics and incidence. Almost all attempts try to fit some model of process to the observed pattern of incidence. They usually succeed: most models can be fit to almost any reasonable pattern. The ease with which different models can be fit to the same data means that one learns relatively little from fitting.

In this book, I advocated two steps to move beyond facilely fitting quantitative models of cellular processes to patterns of incidence. First, breadth of analysis prevents one from uncritically accepting the first quantitative analysis that can be molded to the data. Second, simple comparative hypotheses create the back and forth loop between predictions and tests that reveal causality.

Breast cancer illustrates the importance of breadth in analysis. Breast cancer incidence rises rapidly through midlife and rises slowly after menopause (Figures A.1, A.2); ovarian cancer follows a similar pattern (Figures A.7, A.8). The rate of cell division in female reproductive tissues declines after menopause. It seems natural to relate cell division to incidence, because the rate of mitosis sets one of the major risk factors in cancer. So we may easily fit a model in which mitotic rate shapes the incidence pattern of breast and ovarian cancer.

My broad synthesis of pattern and process in cancer quickly shows how little we learn from the fit of the mitotic rate model to breast and ovarian cancer incidence. On the pattern side, breast and ovarian cancer incidence do follow changes in reproductive status, but so do cancers of the kidney, esophagus, and larynx in both males and females (Figures A.7, A.8).

The broad look at pattern in the Appendix shows that many cancers have rising incidence through midlife followed by a tendency of the incidence curve to flatten (declining acceleration). This common incidence trend of many cancers suggests a universal process.

What sort of universal process might explain declining acceleration later in life? In my theory chapters, I developed a broad conceptual framework for how various processes of progression affect incidence. That broad framework showed that many different processes cause declining acceleration with age.

My favored explanation follows from a universal aspect of multistage progression: as individuals age, they progress stochastically through the early stages of disease. Later in life, they have fewer steps remaining to overt symptoms. With fewer stages remaining, incidence accelerates more slowly with age. This progression scenario fits the data. But I also showed that environmental or genetic heterogeneity can fit the patterns of declining acceleration. By looking broadly at the theory, we avoid latching onto the first good fit.

The theory leaves us with alternative plausible hypotheses, which is all that we should expect from a quantitative framework. But with so many alternatives, some might feel that it is too hard to match biochemical and cellular components to the quantitative processes that drive progression and shape incidence curves.

Perhaps we should wait for all the molecular and cellular details, after which the nature of progression and the final outcome of incidence may be clear. Unfortunately, enumeration will not work. The full list of parts for our plane does not tell us how it flies. Measurements of rate processes by which individual components work locally within the broader system do not solve the problem. To understand cancer, we would certainly like to know how a genetic variant of a DNA repair system alters the somatic mutation rate. But, based on a compilation of such rates, we would not be able to build a large, system-level model that has generality, broad predictive power, and insight into causality. Induction, ever attractive, does not work.

What does work? Simple comparative hypotheses that reveal causality and the design principles that determine outcome: the usual iterative scientific cycle between, on the one hand, the genetic and physiological variations in cells and tissues that define the causes and, on the other hand, the rates at which cancer develops that define the consequences.

Knudson (1971), one of the most cited papers in the history of cancer research, provides a revealing sensor for current trends. Recent citations of Knudson's paper reduce his work to an enumerative slogan and ignore the powerful way in which Knudson himself analyzed causality in cancer. Almost all recent citations of Knudson ascribe to him the "two-hit theory": for many genes, both alleles must be knocked out to cause loss of function and progression toward cancer. However, the two-hit theory was in fact raised several times during the 1960s, before Knudson's publication.

Knudson primarily contributed by figuring out a way to test theories of genetic causation in cancer (see also Ashley 1969a). He compared age-specific incidence curves between inherited and noninherited cases of retinoblastoma. The inherited cases had increased incidence by an amount consistent with an advance of progression by one rate-limiting step. This approach provided a method of analysis by which one could use quantitative comparison of age-specific incidence between two groups to infer underlying processes of progression. In this case, the comparison pointed to a genetic mutation as a key rate-limiting step.

Knudson's approach was simple: predict how a perturbation to process alters outcome. Knudson was particularly successful because he chose to focus on perturbation to heritable properties of cells, in this case, perturbation caused by an inherited mutation, and because he chose to focus on quantitative aspects of the ultimate outcome, the rate at which cancer occurs at different ages.

Current laboratory studies use the same approach. Those lab studies analyze genetic causation by comparing the age-onset curves between different genotypes. If a particular genotype shifts the onset curve to an earlier age, then one ascribes causation to the genetic differences of that genotype relative to the matched control. Those lab studies almost always compare incidence curves in a qualitative way, by simply noting if the incidence curve of a particular subgroup of animals has shifted to an earlier age relative to matched controls. They discard all the quantitative information about outcome contained in the relative rates of progression for different groups.

Throughout this book, I have advocated quantitative comparisons of incidence curves to infer causation. I developed an extensive theoretical framework from which one can predict how genetic or environmental perturbations alter incidence curves. Such comparative predictions can be tested easily in studies of laboratory animals, where the experimenter can control conditions and treatments for different groups.

I have also advocated comparisons of incidence between subgroups of humans. Such comparisons provide particularly interesting information when the subgroups differ in clearly identified aspects of their genetics. Knudson's comparison of inherited and noninherited retinoblastoma provides one example. In that case, identifying the distinct subgroups is relatively easy, because the inherited cases have distinctive patterns of tumor formation when compared to the noninherited cases.

New genomic technologies will soon allow much more refined measurement of genotype in human subjects. With that genetic resolution, one will be able to compare quantitative aspects of age-incidence curves between groups with and without certain genetic attributes. Such comparisons will allow one to ascribe causation to particular genetic differences, and then follow up with analysis of the biochemical processes associated with those genetic differences. This approach demands quantitative evaluation of outcome—the age-incidence curve. I have built the framework required to predict changes in incidence curves based on specific hypotheses about processes of progression.

Appendix:
Incidence

The first section shows plots of cancer incidence for different tissues (Figures A.1–A.12). The second section shows plots of the male:female ratio in incidence for different tissues (Figures A.13–A.18).

Plots of Cancer Incidence at Different Times and Places

The following plots show cancer incidence and acceleration patterns at different time periods and in different countries. In some cases, the acceleration plots fluctuate between countries because of the nature of the data, which may have small numbers of cases at early or late ages. Thus, it is best to focus only on the broad trends in the acceleration plots, particularly those patterns that recur in different years and in different locations. For example, prostate cancer shows a remarkably strong and linear decline in acceleration beginning in midlife (Figure A.2). Some cancers show midlife peaks in acceleration, for example, colon and bladder cancer (Figure A.4).

Cervical cancer has an acceleration close to zero throughout life, with higher fluctuations outside the USA probably caused by smaller samples for those other countries (Figure A.12). However, cervical cancer in the USA follows different patterns of acceleration in different ethnic groups (not shown), emphasizing that external factors such as environment and lifestyle can strongly affect incidence and acceleration. Given the variability in potential causal factors, the data in the following plots can be used only to suggest possible hypotheses for further study.

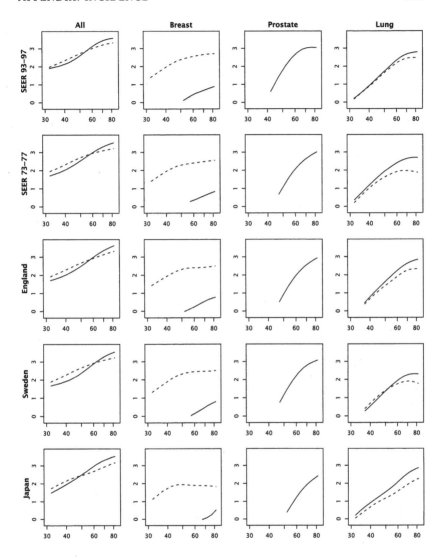

Figure A.1   Age-specific incidence for different time periods and geographic locations. Male cases shown by solid lines; female cases shown by dashed lines. The different databases are: SEER 93–97 and SEER 73–77 from the SEER database (http://seer.cancer.gov/) in the USA for 1993–1997 and 1973–1977 using white individuals in the standard nine registries that have been in use since 1973; England, Sweden, and Japan from the CI5 database (Parkin et al. 2002) for 1993–1997 (for Japan, I excluded the Hiroshima registry, which had data for a different range of dates).

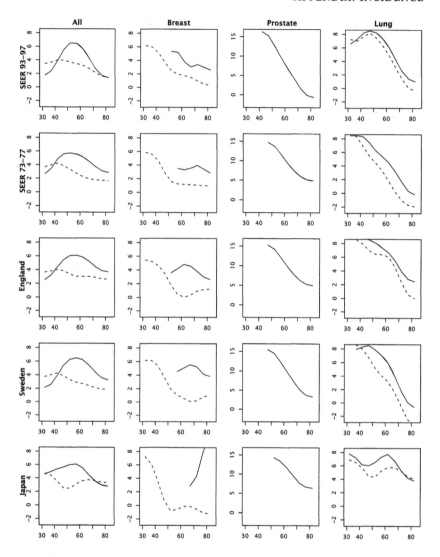

Figure A.2   Age-specific acceleration for different time periods and geographic locations. Male cases shown by solid lines; female cases shown by dashed lines. Data description as in Figure A.1. The prostate acceleration is shown on a different scale, to accomodate the very high acceleration that occurs in midlife.

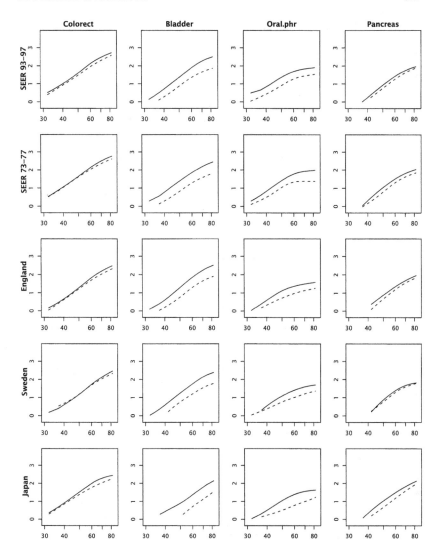

Figure A.3 Age-specific incidence for different time periods and geographic locations. Male cases shown by solid lines; female cases shown by dashed lines. Data description as in Figure A.1. SEER plots show combined data for colon and rectal cancer, other countries show colon cancer only. Colon cancer is more common than rectal cancer, so these plots are roughly comparable.

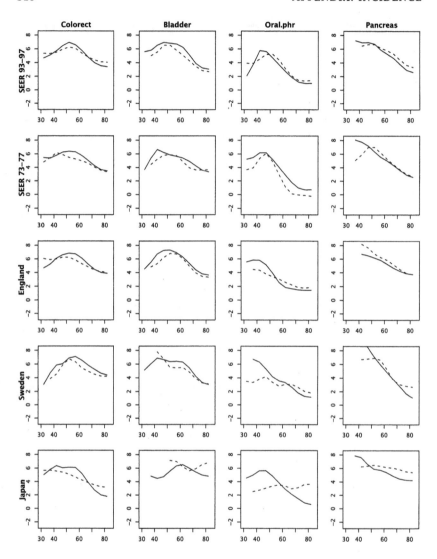

Figure A.4   Age-specific acceleration for different time periods and geographic locations. Male cases shown by solid lines; female cases shown by dashed lines. Data description as in Figure A.1. SEER plots show combined data for colon and rectal cancer, other countries show colon cancer only. Colon cancer is more common than rectal cancer, so these plots are roughly comparable.

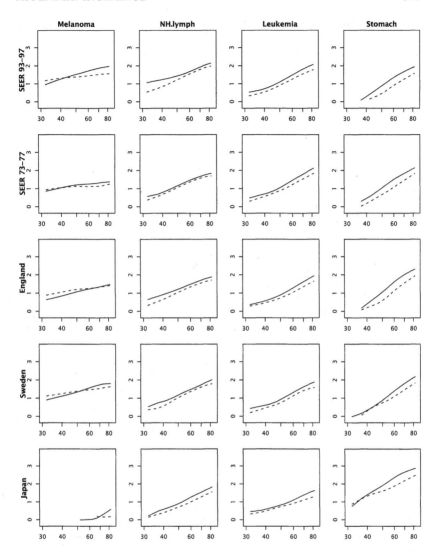

Figure A.5 Age-specific incidence for different time periods and geographic locations. Male cases shown by solid lines; female cases shown by dashed lines. Data description as in Figure A.1.

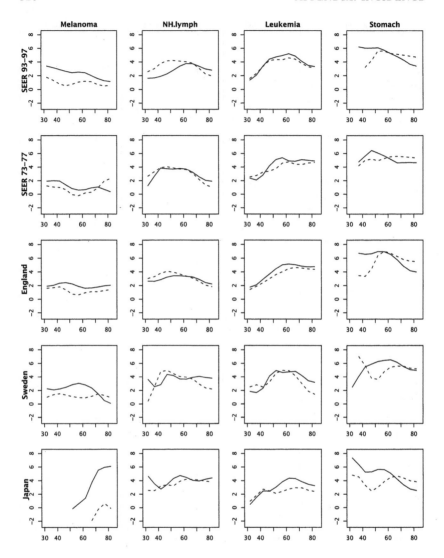

Figure A.6   Age-specific acceleration for different time periods and geographic locations. Male cases shown by solid lines; female cases shown by dashed lines. Data description as in Figure A.1.

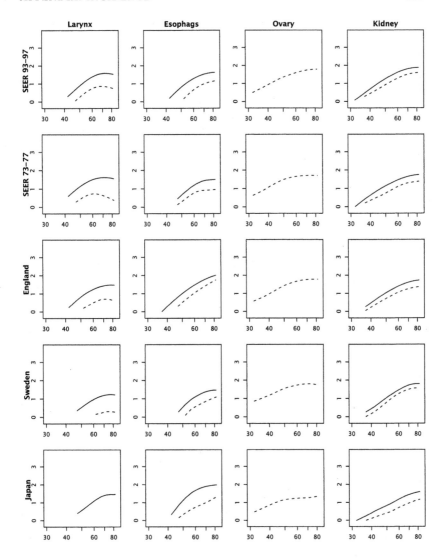

Figure A.7   Age-specific incidence for different time periods and geographic locations. Male cases shown by solid lines; female cases shown by dashed lines. Data description as in Figure A.1.

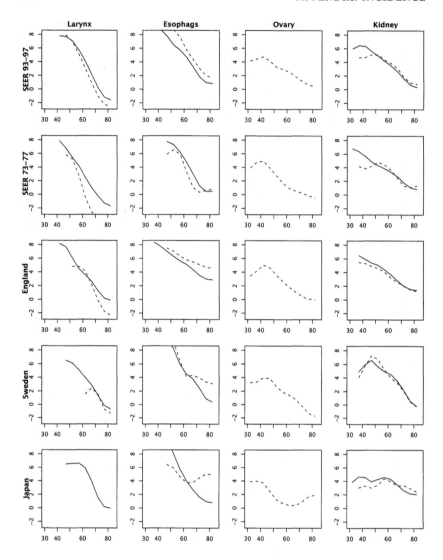

Figure A.8   Age-specific acceleration for different time periods and geographic locations. Male cases shown by solid lines; female cases shown by dashed lines. Data description as in Figure A.1.

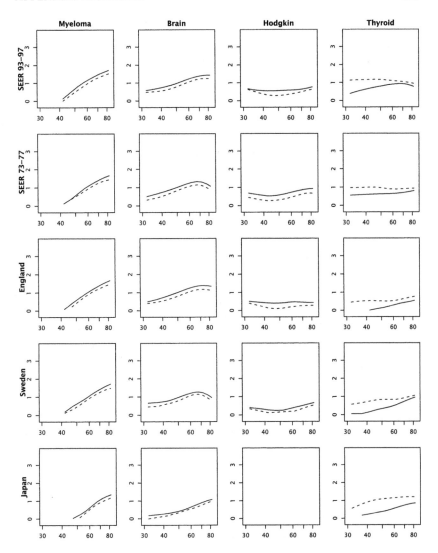

Figure A.9   Age-specific incidence for different time periods and geographic locations. Male cases shown by solid lines; female cases shown by dashed lines. Data description as in Figure A.1.

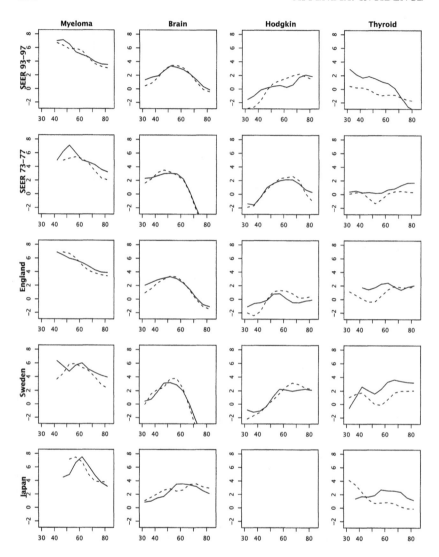

Figure A.10 Age-specific acceleration for different time periods and geographic locations. Male cases shown by solid lines; female cases shown by dashed lines. Data description as in Figure A.1.

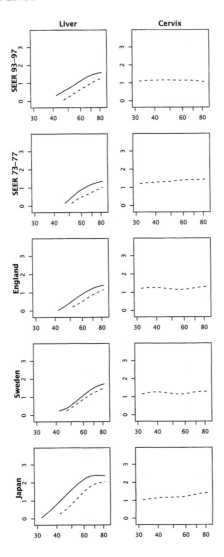

Figure A.11   Age-specific incidence for different time periods and geographic locations. Male cases shown by solid lines; female cases shown by dashed lines. Data description as in Figure A.1.

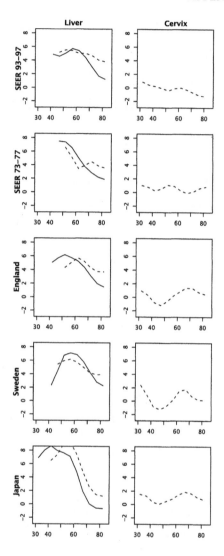

Figure A.12 Age-specific acceleration for different time periods and geographic locations. Male cases shown by solid lines; female cases shown by dashed lines. Data description as in Figure A.1.

Sex Differences in Incidence

Figures A.13–A.18 show the male:female ratios for the major adult cancers. The plots highlight two kinds of information. First, the values on the y axis measure the male:female ratio, with positive values for male excess and negative values for female excess. The scaling is explained in the legend of Figure A.13. Second, the trend in each plot shows the relative acceleration of male and female incidence with age. For example, in Figure A.13, the positive trend for lung cancer shows that male incidence accelerates with age more rapidly than does female incidence, probably because males have smoked more than females, at least in the past. Positive trends also occur consistently for the colon, bladder, melanoma, leukemia, and thyroid. Negative trends may occur for the pancreas, esophagus, and liver, but the results for those tissues are mixed among locations. Simple nonlinear curves seem to explain the patterns for the stomach and Hodgkin's, and maybe also for oral-pharyngeal cancers.

The patterns of relative male:female incidence probably arise from differences between males and females in exposure to carcinogens, to expression of different hormone profiles, or from different patterns of tissue growth, damage, or repair. At present, the observed patterns serve mainly to guide the development of hypotheses along these lines.

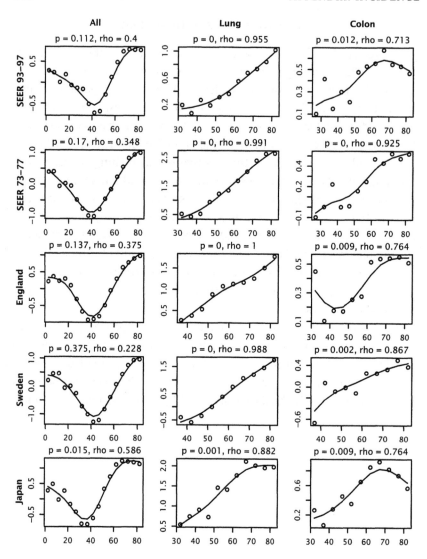

Figure A.13 Ratio of male to female age-specific incidence. The y axis shows male incidence rate divided by female incidence rate for each age, given on a $\log_2$ scale. This scaling maps an equal male:female incidence ratio to a value of zero; each unit on the scale means a two-fold change in relative incidence, with negative values occurring when female incidence exceeds male incidence. Each plot shows the Spearman's rho correlation coefficient and p-value; a p-value of zero means $p < 0.0005$. Positive correlations occur when there is an increasing trend in the ratio of male to female incidence with increasing age. Note that the scales differ between plots, using the maximum range of the data to emphasize the shapes of the curves. The data are the same as used in Figures A.1–A.11.

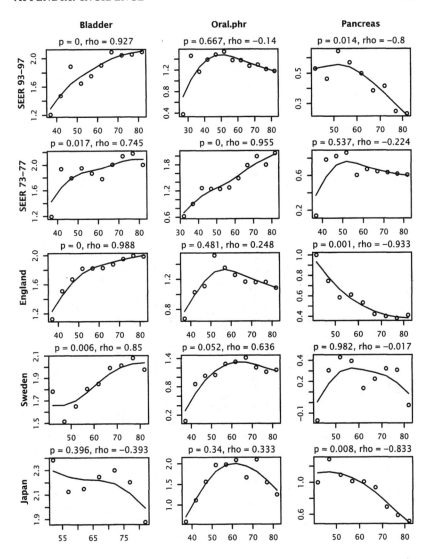

Figure A.14   Sex differences in incidence, as in Figure A.13.

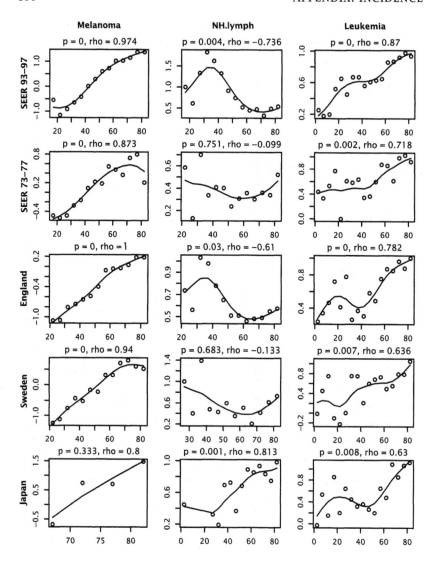

Figure A.15   Sex differences in incidence, as in Figure A.13.

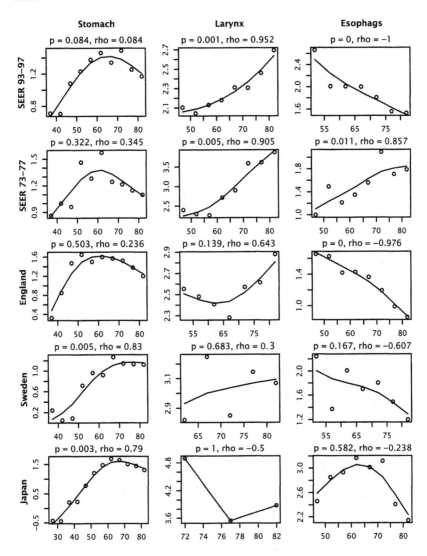

Figure A.16   Sex differences in incidence, as in Figure A.13.

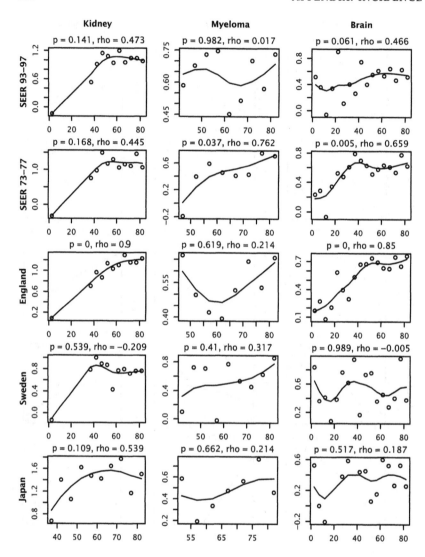

Figure A.17   Sex differences in incidence, as in Figure A.13.

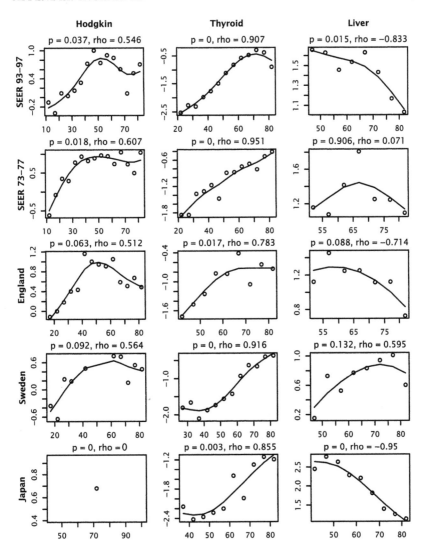

Figure A.18   Sex differences in incidence, as in Figure A.13.

References

Adami, H. O., Hunter, D., and Trichopoulos, D., eds. 2002. *Textbook of Cancer Epidemiology*, Vol. 33 of *Monographs in Epidemiology and Biostatistics*. Oxford University Press, New York.

Ames, B. N., and Gold, L. S. 1990. Too many rodent carcinogens: mitogenesis increases mutagenesis. *Science* 249:970-971.

Anderson, D. E. 1970. Genetic varieties of neoplasia. In *Genetic Concepts and Neoplasia: Proceedings of the 23rd Symposium on Fundamental Cancer Research*, pp. 85-104. Williams & Wilkins, Baltimore.

Andreassi, M. G., Botto, N., Colombo, M. G., Biagini, A., and Clerico, A. 2000. Genetic instability and atherosclerosis: can somatic mutations account for the development of cardiovascular diseases? *Environmental and Molecular Mutagenesis* 35:265-269.

Anglian Breast Cancer Study Group 2000. Prevalence and penetrance of *BRCA1* and *BRCA2* mutations in a population-based series of breast cancer cases. *British Journal of Cancer* 83:1301-1308.

Armakolas, A., and Klar, A. J. 2006. Cell type regulates selective segregation of mouse chromosome 7 DNA strands in mitosis. *Science* 311:1146-1149.

Armitage, P. 1953. A note on the time-homogeneous birth process. *Journal of the Royal Statistical Society, Series B (Methodological)* 15:90-91.

Armitage, P., and Doll, R. 1954. The age distribution of cancer and a multi-stage theory of carcinogenesis. *British Journal of Cancer* 8:1-12.

Armitage, P., and Doll, R. 1957. A two-stage theory of carcinogenesis in relation to the age distribution of human cancer. *British Journal of Cancer* 11:161-169.

Armitage, P., and Doll, R. 1961. Stochastic models for carcinogenesis. In Neyman, J., ed., *Proceedings of the Fourth Berkeley Symposium on Mathematical Statistics and Probability*, pp. 19-38. University of California Press, Berkeley.

Armstrong, B., and Doll, R. 1975. Environmental factors and cancer incidence and mortality in different countries, with special reference to dietary practices. *International Journal of Cancer* 15:617-631.

Ashley, D. J. 1969a. Colonic cancer arising in polyposis coli. *Journal of Medical Genetics* 6:376-378.

Ashley, D. J. 1969b. The two "hit" and multiple "hit" theories of carcinogenesis. *British Journal of Cancer* 23:313-328.

Ayabe, T., Satchell, D. P., Wilson, C. L., Parks, W. C., Selsted, M. E., and Ouellette, A. J. 2000. Secretion of microbicidal alpha-defensins by intestinal Paneth cells in response to bacteria. *Nature Immunology* 1:113-118.

Bach, S. P., Renehan, A. G., and Potten, C. S. 2000. Stem cells: the intestinal stem cell as a paradigm. *Carcinogenesis* 21:469-476.

Backvall, H., Asplund, A., Gustafsson, A., Sivertsson, A., Lundeberg, J., and Ponten, F. 2005. Genetic tumor archeology: microdissection and genetic heterogeneity in squamous and basal cell carcinoma. *Mutation Research/Fundamental and Molecular Mechanisms of Mutagenesis* 571:65-79.

Bahar, R., Hartmann, C. H., Rodriguez, K. A., Denny, A. D., Busuttil, R. A., Dolle, M. E., Calder, R. B., Chisholm, G. B., Pollock, B. H., Klein, C. A., and Vijg, J. 2006. Increased cell-to-cell variation in gene expression in ageing mouse heart. *Nature* 441:1011-1014.

Bapat, S. A. 2006. Evolution of cancer stem cells. *Seminars in Cancer Biology* [doi:10.1016/j.semcancer.2006.05.001].

Barbacid, M. 1987. *ras* genes. *Annual Review of Biochemistry* 56:779-827.

Barton, N. H., and Keightley, P. D. 2002. Understanding quantitative genetic variation. *Nature Reviews Genetics* 3:11-21.

Beckman, R. A., and Loeb, L. A. 2005. Genetic instability in cancer: theory and experiment. *Seminars in Cancer Biology* 15:423-435.

Berenblum, I. 1941. The cocarcinogenic action of croton resin. *Cancer Research* 1:44-48.

Berenblum, I., and Shubik, P. 1947a. A new, quantitative approach to the study of stages of chemical carcinogenesis in the mouse's skin. *British Journal of Cancer* 1:383-391.

Berenblum, I., and Shubik, P. 1947b. The role of croton oil applications associated with a single painting of a carcinogen in tumour induction of the mouse's skin. *British Journal of Cancer* 1:379-382.

Berenblum, I., and Shubik, P. 1949. The persistence of latent tumour cells induced in the mouse's skin by a single application of 9:10-dimethyl-1:2-benzanthracene. *British Journal of Cancer* 3:384-386.

Bernstein, C., Bernstein, H., Payne, C. M., and Garewal, H. 2002. DNA repair/proapoptotic dual-role proteins in five major DNA repair pathways: fail-safe protection against carcinogenesis. *Mutation Research* 511:145-178.

Berwick, M., and Vineis, P. 2000. Markers of DNA repair and susceptibility to cancer in humans: an epidemiologic review. *Journal of the National Cancer Institute* 92:874-897.

Boland, C. R. 2002. Heredity nonpolyposis colorectal cancer (HNPCC). In Vogelstein, B., and Kinzler, K. W., eds., *The Genetic Basis of Human Cancer* (2nd edition)., pp. 307-321. McGraw-Hill, New York.

Bond, G. L., Hu, W., Bond, E. E., Robins, H., Lutzker, S. G., Arva, N. C., Bargonetti, J., Bartel, F., Taubert, H., Wuerl, P., Onel, K., Yip, L., Hwang, S. J., Strong, L. C., Lozano, G., and Levine, A. J. 2004. A single nucleotide polymorphism in the *MDM2* promoter attenuates the p53 tumor suppressor pathway and accelerates tumor formation in humans. *Cell* 119:591-602.

Bonsing, B. A., Corver, W. E., Fleuren, G. J., Cleton-Jansen, A. M., Devilee, P., and Cornelisse, C. J. 2000. Allelotype analysis of flow-sorted breast cancer cells demonstrates genetically related diploid and aneuploid subpopulations in primary tumors and lymph node metastases. *Genes, Chromosomes and Cancer* 28:173-183.

Boon, L. M., Brouillard, P., Irrthum, A., Karttunen, L., Warman, M. L., Rudolph, R., Mulliken, J. B., Olsen, B. R., and Vikkula, M. 1999. A gene for inherited cutaneous venous anomalies ("glomangiomas") localizes to chromosome 1p21-22. *American Journal of Human Genetics* 65:125-133.

Boveri, T. 1914. *Zur Frage der Entstehung maligner Tumoren.* Fischer, Jena.

Boveri, T. 1929. *The Origin of Malignant Tumors.* Williams and Wilkins, Baltimore.

Braakhuis, B. J., Leemans, C. R., and Brakenhoff, R. H. 2005. Expanding fields of genetically altered cells in head and neck squamous carcinogenesis. *Seminars in Cancer Biology* 15:113-120.

Braakhuis, B. J., Tabor, M. P., Kummer, J. A., Leemans, C. R., and Brakenhoff, R. H. 2003. A genetic explanation of Slaughter's concept of field cancerization: evidence and clinical implications. *Cancer Research* 63:1727-1730.

Bradford, G. B., Williams, B., Rossi, R., and Bertoncello, I. 1997. Quiescence, cycling, and turnover in the primitive hematopoietic stem cell compartment. *Experimental Hematology* 25:445-453.

Brash, D. E. 2006. Roles of the transcription factor p53 in keratinocyte carcinomas. *British Journal of Dermatology* 154 (Suppl 1):8-10.

Breivik, J., and Gaudernack, G. 1999a. Carcinogenesis and natural selection: a new perspective to the genetics and epigenetics of colorectal cancer. *Advances in Cancer Research* 76:187-212.

Breivik, J., and Gaudernack, G. 1999b. Genomic instability, DNA methylation, and natural selection in colorectal carcinogenesis. *Seminars in Cancer Biology* 9:245-254.

Brouillard, P., Boon, L. M., Mulliken, J. B., Enjolras, O., Ghassibe, M., Warman, M. L., Tan, O. T., Olsen, B. R., and Vikkula, M. 2002. Mutations in a novel factor, glomulin, are responsible for glomuvenous malformations ("glomangiomas"). *American Journal of Human Genetics* 70:866-874.

Brouillard, P., Ghassibe, M., Penington, A., Boon, L. M., Dompmartin, A., Temple, I. K., Cordisco, M., Adams, D., Piette, F., Harper, J. I., Syed, S., Boralevi, F., Taieb, A., Danda, S., Baselga, E., Enjolras, O., Mulliken, J. B., and Vikkula, M. 2005. Four common glomulin mutations cause two thirds of glomuvenous malformations ("familial glomangiomas"): evidence for a founder effect. *Journal of Medical Genetics* 42:e13.

Brouillard, P., and Vikkula, M. 2003. Vascular malformations: localized defects in vascular morphogenesis. *Clinical Genetics* 63:340-351.

Brown, C. C., and Chu, K. C. 1987. Use of multistage models to infer stage affected by carcinogenic exposure: example of lung cancer and cigarette smoking. *Journal of Chronic Diseases* 40 (Suppl 2):171S-179S.

Brown, K., Buchmann, A., and Balmain, A. 1990. Carcinogen-induced mutations in the mouse *c-Ha-ras* gene provide evidence of multiple pathways for tumor progression. *Proceedings of the National Academy of Sciences of the United States of America* 87:538-542.

Brown, K., Burns, P. A., and Balmain, A. 1995. Transgenic approaches to understanding the mechanisms of chemical carcinogenesis in mouse skin. *Toxicology Letters* 82-83:123-130.

Buermeyer, A. B., Deschenes, S. M., Baker, S. M., and Liskay, R. M. 1999. Mammalian DNA mismatch repair. *Annual Review of Genetics* 33:533-564.

Burch, P. R. 1963. Human cancer: Mendelian inheritance or vertical transmission? *Nature* 197:1042-1045.

Burch, P. R. 1964. Genetic carrier frequency for lung cancer. *Nature* 202:711-712.

Burdette, W. J. 1955. The significance of mutation in relation to the origin of tumors: a review. *Cancer Research* 15:201-226.

Burns, P. A., Kemp, C. J., Gannon, J. V., Lane, D. P., Bremner, R., and Balmain, A. 1991. Loss of heterozygosity and mutational alterations of the *p53* gene in skin tumours of interspecific hybrid mice. *Oncogene* 6:2363-2369.

Buss, L. W. 1987. *The Evolution of Individuality.* Princeton University Press, Princeton, NJ.

Cairns, J. 1975. Mutation selection and the natural history of cancer. *Nature* 255:197-200.

Cairns, J. 1978. *Cancer: Science and Society.* W. H. Freeman, San Francisco.

Cairns, J. 1997. *Matters of Life and Death.* Princeton University Press, Princeton, NJ.

Cairns, J. 1998. Mutation and cancer: the antecedents to our studies of adaptive mutation. *Genetics* 148:1433-1440.

Cairns, J. 2002. Somatic stem cells and the kinetics of mutagenesis and carcinogenesis. *Proceedings of the National Academy of Sciences of the United States of America* 99:10567-10570.

Calabrese, P., Tavare, S., and Shibata, D. 2004. Pretumor progression: clonal evolution of human stem cell populations. *American Journal of Pathology* 164:1337-1346.

Carey, J. R. 2003. *Longevity: The Biology of Life Span.* Princeton University Press, Princeton, NJ.

Charles, D. R., and Luce-Clausen, E. M. 1942. The kinetics of papilloma formation in benzpyrene-treated mice. *Cancer Research* 2:261-263.

Charlesworth, B., and Partridge, L. 1997. Ageing: levelling of the grim reaper. *Current Biology* 7:R440-442.

Chen, P. C., Dudley, S., Hagen, W., Dizon, D., Paxton, L., Reichow, D., Yoon, S. R., Yang, K., Arnheim, N., Liskay, R. M., and Lipkin, S. M. 2005. Contributions by *MutL* homologues *Mlh3* and *Pms2* to DNA mismatch repair and tumor suppression in the mouse. *Cancer Research* 65:8662-8670.

Cheng, T. C., Chen, S. T., Huang, C. S., Fu, Y. P., Yu, J. C., Cheng, C. W., Wu, P. E., and Shen, C. Y. 2005. Breast cancer risk associated with genotype polymorphism of the catechol estrogen-metabolizing genes: a multigenic study on cancer susceptibility. *International Journal of Cancer* 113:345–353.

Cheshier, S. H., Morrison, S. J., Liao, X., and Weissman, I. L. 1999. In vivo proliferation and cell cycle kinetics of long-term self-renewing hematopoietic stem cells. *Proceedings of the National Academy of Sciences of the United States of America* 96:3120–3125.

Chuong, C. M., Dhouailly, D., Gilmore, S., Forest, L., Shelley, W. B., Stenn, K. S., Maini, P., Michon, F., Parimoo, S., Cadau, S., Demongeot, J., Zheng, Y., Paus, R., and Happle, R. 2006. What is the biological basis of pattern formation of skin lesions? *Experimental Dermatology* 15:547–549.

Clara, M., Herschel, K., and Ferner, H. 1974. *Atlas of Normal Microscopic Anatomy of Man.* Urban and Schwarzenberg, New York.

Clarke, R. B., Anderson, E., Howell, A., and Potten, C. S. 2003. Regulation of human breast epithelial stem cells. *Cell Proliferation* 36 (Suppl 1):45–58.

Clemmesen, J. 1964. *Statistical Studies in the Aetiology of Malignant Neoplasms,* Vol. 174 of *Acta Pathologica et Microbiologica Scandinavica. Supplement.* Munksgaard, Copenhagen.

Clemmesen, J. 1969. *Statistical Studies in the Aetiology of Malignant Neoplasms,* Vol. 209 of *Acta Pathologica et Microbiologica Scandinavica. Supplement.* Munksgaard, Copenhagen.

Clemmesen, J. 1974. *Statistical Studies in the Aetiology of Malignant Neoplasms,* Vol. 247 of *Acta Pathologica et Microbiologica Scandinavica. Supplement.* Munksgaard, Copenhagen.

Cloos, J., Nieuwenhuis, E. J., Boomsma, D. I., Kuik, D. J., van der Sterre, M. L., Arwert, F., Snow, G. B., and Braakhuis, B. J. 1999. Inherited susceptibility to bleomycin-induced chromatid breaks in cultured peripheral blood lymphocytes. *Journal of the National Cancer Institute* 91:1125–1130.

Collaborative Group on Hormonal Factors in Breast Cancer 2001. Familial breast cancer: collaborative reanalysis of individual data from 52 epidemiological studies including 58,209 women with breast cancer and 101,986 women without the disease. *Lancet* 358:1389–1399.

Cook, P. J., Doll, R., and Fellingham, S. A. 1969. A mathematical model for the age distribution of cancer in man. *International Journal of Cancer* 4:93–112.

Cotsarelis, G., Sun, T. T., and Lavker, R. M. 1990. Label-retaining cells reside in the bulge area of pilosebaceous unit: implications for follicular stem cells, hair cycle, and skin carcinogenesis. *Cell* 61:1329–1337.

Couch, F. J., and Weber, B. L. 1996. Mutations and polymorphisms in the familial early-onset breast cancer (*BRCA1*) gene. Breast Cancer Information Core. *Human Mutation* 8:8–18.

Couch, F. J., and Weber, B. L. 2002. Breast cancer. In Vogelstein, B., and Kinzler, K. W., eds., *The Genetic Basis of Human Cancer* (2nd edition)., pp. 549–581. McGraw-Hill, New York.

Crowe, F. W., Schull, W. J., and Neel, J. V. 1956. *A Clinical, Pathological and*

Genetic Study of Multiple Neurofibromatosis. Charles C Thomas, Springfield, IL.

Cunningham, M. L., and Matthews, H. B. 1995. Cell proliferation as a determining factor for the carcinogenicity of chemicals: studies with mutagenic carcinogens and mutagenic noncarcinogens. *Toxicology Letters* 82-83:9-14.

Czene, K., Lichtenstein, P., and Hemminki, K. 2002. Environmental and heritable causes of cancer among 9.6 million individuals in the Swedish Family-Cancer Database. *International Journal of Cancer* 99:260-266.

Dahmen, R. P., Koch, A., Denkhaus, D., Tonn, J. C., Sorensen, N., Berthold, F., Behrens, J., Birchmeier, W., Wiestler, O. D., and Pietsch, T. 2001. Deletions of *AXIN1*, a component of the WNT/wingless pathway, in sporadic medulloblastomas. *Cancer Research* 61:7039-7043.

Day, N. E., and Brown, C. C. 1980. Multistage models and primary prevention of cancer. *Journal of the National Cancer Institute* 64:977-989.

de Boer, J. G. 2002. Polymorphisms in DNA repair and environmental interactions. *Mutation Research* 509:201-210.

de la Chapelle, A. 2004. Genetic predisposition to colorectal cancer. *Nature Reviews Cancer* 4:769-780.

de Rooij, D. G. 1998. Stem cells in the testis. *International Journal of Experimental Pathology* 79:67-80.

Dean, M., Fojo, T., and Bates, S. 2005. Tumour stem cells and drug resistance. *Nature Reviews Cancer* 5:275-284.

Deelman, H. T. 1927. The part played by injury and repair in the development of cancer, with remarks on the growth of experimental cancer. *British Medical Journal* 1:872-874.

DeMars, R. 1970. Discussion comments following a paper by D. E. Anderson. In *Genetic Concepts and Neoplasia: Proceedings of the 23rd Symposium on Fundamental Cancer Research*, pp. 105-106. Williams & Wilkins, Baltimore.

Doll, R. 1971. The age distribution of cancer: implications for models of carcinogenesis. *Journal of the Royal Statistical Society, Series A* 134:133-166.

Doll, R. 1998. Uncovering the effects of smoking: historical perspective. *Statistical Methods in Medical Research* 7:87-117.

Doll, R., and Peto, R. 1978. Cigarette smoking and bronchial carcinoma: dose and time relationships among regular smokers and lifelong non-smokers. *Journal of Epidemiology and Community Health* 32:303-313.

Dontu, G., Al-Hajj, M., Abdallah, W. M., Clarke, M. F., and Wicha, M. S. 2003. Stem cells in normal breast development and breast cancer. *Cell Proliferation* 36 (Suppl 1):59-72.

Douma, S., Van Laar, T., Zevenhoven, J., Meuwissen, R., Van Garderen, E., and Peeper, D. S. 2004. Suppression of anoikis and induction of metastasis by the neurotrophic receptor TrkB. *Nature* 430:1034-1039.

Drake, J. W., Charlesworth, B., Charlesworth, D., and Crow, J. F. 1998. Rates of spontaneous mutation. *Genetics* 148:1667-1686.

Druckrey, H. 1967. Quantitative aspects in chemical carcinogenesis. In Truhaut, R., ed., *Potential Carcinogenic Hazards from Drugs*, pp. 60–78. Springer-Verlag, Berlin.

Dyson, F. 2004. A meeting with Enrico Fermi. *Nature* 427:297.

Edelmann, L., and Edelmann, W. 2004. Loss of DNA mismatch repair function and cancer predisposition in the mouse: animal models for human hereditary nonpolyposis colorectal cancer. *American Journal of Medical Genetics. Part C, Seminars in Medical Genetics* 129:91–99.

Egger, G., Liang, G., Aparicio, A., and Jones, P. A. 2004. Epigenetics in human disease and prospects for epigenetic therapy. *Nature* 429:457–463.

Evans, H. J. 1984. Genetic damage and cancer. In Bishop, J. M., Rowley, J. D., and Greaves, M., eds., *Genes and Cancer*, pp. 3–18. Alan R. Liss, New York.

Fearnhead, N. S., Wilding, J. L., Winney, B., Tonks, S., Bartlett, S., Bicknell, D. C., Tomlinson, I. P., Mortensen, N. J., and Bodmer, W. F. 2004. Multiple rare variants in different genes account for multifactorial inherited susceptibility to colorectal adenomas. *Proceedings of the National Academy of Sciences of the United States of America* 101:15992–15997.

Fearon, E. R. 2002. Tumor-suppressor genes. In Vogelstein, B., and Kinzler, K. W., eds., *The Genetic Basis of Human Cancer* (2nd edition)., pp. 197–206. McGraw-Hill, New York.

Fearon, E. R., and Vogelstein, B. 1990. A genetic model for colorectal tumorigenesis. *Cell* 61:759–767.

Feinberg, A. P., and Tycko, B. 2004. The history of cancer epigenetics. *Nature Reviews Cancer* 4:143–153.

Feldser, D. M., Hackett, J. A., and Greider, C. W. 2003. Telomere dysfunction and the initiation of genome instability. *Nature Reviews Cancer* 3:623–627.

Felsenstein, J. 2003. *Inferring Phylogenies*. Sinauer Associates, Sunderland, MA.

Fidler, I. J. 2003. The pathogenesis of cancer metastasis: the "seed and soil" hypothesis revisited. *Nature Reviews Cancer* 3:453–458.

Fishel, R. 2001. The selection for mismatch repair defects in hereditary nonpolyposis colorectal cancer: revising the mutator hypothesis. *Cancer Research* 61:7369–7374.

Fisher, J. C. 1958. Multiple-mutation theory of carcinogenesis. *Nature* 181:651–652.

Fisher, J. C., and Hollomon, J. H. 1951. A hypothesis for the origin of cancer foci. *Cancer* 4:916–918.

Folkman, J. 2002. Role of angiogenesis in tumor growth and metastasis. *Seminars in Oncology* 29:15–18.

Folkman, J. 2003. Fundamental concepts of the angiogenic process. *Current Molecular Medicine* 3:643–651.

Forbes, W. F., and Gibberd, R. W. 1984. Mathematical models of carcinogenesis: a review. *Mathematical Scientist* 9:95–110.

Ford, D., Easton, D. F., Stratton, M. R., Narod, S., Goldgar, D., Devilee, P., Bishop,

D. T., Weber, B. L., Lenoir, G., Chang-Claude, J., Sobol, H., Teare, M. D., Struewing, J. P., Arason, A., Scherneck, S., Peto, J., Rebbeck, T. R., Tonin, P., Neuhausen, S., Barkardottir, R., Eyfjord, J., Lynch, H. T., Ponder, B. A. et al. 1998. Genetic heterogeneity and penetrance analysis of the *BRCA1* and *BRCA2* genes in breast cancer families. The Breast Cancer Linkage Consortium. *American Journal of Human Genetics* 62:676–689.

Foulds, L. 1969. *Neoplastic Development*, Vol. 1. Academic Press, New York.

Fraga, M. F., Ballestar, E., Paz, M. F., Ropero, S., Setien, F., Ballestar, M. L., Heine-Suner, D., Cigudosa, J. C., Urioste, M., Benitez, J., Boix-Chornet, M., Sanchez-Aguilera, A., Ling, C., Carlsson, E., Poulsen, P., Vaag, A., Stephan, Z., Spector, T. D., Wu, Y.-Z., Plass, C., and Esteller, M. 2005. Epigenetic differences arise during the lifetime of monozygotic twins. *Proceedings of the National Academy of Sciences of the United States of America* 102:10604–10609.

Frame, S., Crombie, R., Liddell, J., Stuart, D., Linardopoulos, S., Nagase, H., Portella, G., Brown, K., Street, A., Akhurst, R., and Balmain, A. 1998. Epithelial carcinogenesis in the mouse: correlating the genetics and the biology. *Philosophical Transactions of the Royal Society of London. Series B: Biological Sciences* 353:839–845.

Frank, S. A. 1995. Mutual policing and repression of competition in the evolution of cooperative groups. *Nature* 377:520–522.

Frank, S. A. 2003a. Perspective: repression of competition and the evolution of cooperation. *Evolution* 57:693–705.

Frank, S. A. 2003b. Somatic mosaicism and cancer: inference based on a conditional Luria-Delbruck distribution. *Journal of Theoretical Biology* 223:405–412.

Frank, S. A. 2003c. Somatic mutation: early cancer steps depend on tissue architecture. *Current Biology* 13:R261–263.

Frank, S. A. 2004a. A multistage theory of age-specific acceleration in human mortality. *BMC Biology* 2:16.

Frank, S. A. 2004b. Age-specific acceleration of cancer. *Current Biology* 14:242–246.

Frank, S. A. 2004c. Commentary: Mathematical models of cancer progression and epidemiology in the age of high throughput genomics. *International Journal of Epidemiology* 33:1179–1181.

Frank, S. A. 2004d. Genetic predisposition to cancer—insights from population genetics. *Nature Reviews Genetics* 5:764–772.

Frank, S. A. 2004e. Genetic variation in cancer predisposition: mutational decay of a robust genetic control network. *Proceedings of the National Academy of Sciences of the United States of America* 101:8061–8065.

Frank, S. A. 2005. Age-specific incidence of inherited versus sporadic cancers: a test of the multistage theory of carcinogenesis. *Proceedings of the National Academy of Sciences of the United States of America* 102:1071–1075.

Frank, S. A., Chen, P. C., and Lipkin, S. M. 2005. Kinetics of cancer: a method to test hypotheses of genetic causation. *BMC Cancer* 5:163.

Frank, S. A., Iwasa, Y., and Nowak, M. A. 2003. Patterns of cell division and the risk of cancer. *Genetics* 163:1527-1532.

Frank, S. A., and Nowak, M. A. 2003. Developmental predisposition to cancer. *Nature* 422:494.

Frank, S. A., and Nowak, M. A. 2004. Problems of somatic mutation and cancer. *Bioessays* 26:291-299.

Frayling, I. M., Beck, N. E., Ilyas, M., Dove-Edwin, I., Goodman, P., Pack, K., Bell, J. A., Williams, C. B., Hodgson, S. V., Thomas, H. J., Talbot, I. C., Bodmer, W. F., and Tomlinson, I. P. 1998. The *APC* variants I1307K and E1317Q are associated with colorectal tumors, but not always with a family history. *Proceedings of the National Academy of Sciences of the United States of America* 95:10722-10727.

Freedman, D. A., and Navidi, W. C. 1989. Multistage models for carcinogenesis. *Environmental Health Perspectives* 81:169-188.

Friedewald, W. F., and Rous, P. 1944. The initiating and promoting elements in tumor production: an analysis of the effects of tar, benzopyrene and methylcholanthrene on rabbit skin. *Journal of Experimental Medicine* 80:101-125.

Frumkin, D., Wasserstrom, A., Kaplan, S., Feige, U., and Shapiro, E. 2005. Genomic variability within an organism exposes its cell lineage tree. *PLoS Computational Biology* 1:e50.

Gaffney, M., and Altshuler, B. 1988. Examination of the role of cigarette smoke in lung carcinogenesis using multistage models. *Journal of the National Cancer Institute* 80:925-931.

Garcia-Closas, M., Malats, N., Real, F. X., Welch, R., Kogevinas, M., Chatterjee, N., Pfeiffer, R., Silverman, D., Dosemeci, M., Tardon, A., Serra, C., Carrato, A., Garcia-Closas, R., Castano-Vinyals, G., Chanock, S., Yeager, M., and Rothman, N. 2006. Genetic variation in the nucleotide excision repair pathway and bladder cancer risk. *Cancer Epidemiology, Biomarkers and Prevention* 15:536-542.

Gavrilov, L. A., and Gavrilova, N. S. 2001. The reliability theory of aging and longevity. *Journal of Theoretical Biology* 213:527-545.

Genereux, D. P., Miner, B. E., Bergstrom, C. T., and Laird, C. D. 2005. A population-epigenetic model to infer site-specific methylation rates from double-stranded DNA methylation patterns. *Proceedings of the National Academy of Sciences of the United States of America* 102:5802-5807.

Ghazizadeh, S., and Taichman, L. B. 2001. Multiple classes of stem cells in cutaneous epithelium: a lineage analysis of adult mouse skin. *EMBO Journal* 20:1215-1222.

Ghazizadeh, S., and Taichman, L. B. 2005. Organization of stem cells and their progeny in human epidermis. *Journal of Investigative Dermatology* 124:367–372.

Giebel, B., Zhang, T., Beckmann, J., Spanholtz, J., Wernet, P., Ho, A. D., and Punzel, M. 2006. Primitive human hematopoietic cells give rise to differentially specified daughter cells upon their initial cell division. *Blood* 107:2146–2152.

Gottlieb, B., Beitel, L. K., and Trifiro, M. A. 2001. Somatic mosaicism and variable expressivity. *Trends in Genetics* 17:79–82.

Greenhalgh, D. A., Wang, X. J., Donehower, L. A., and Roop, D. R. 1996. Paradoxical tumor inhibitory effect of *p53* loss in transgenic mice expressing epidermal-targeted *v-rasHa, v-fos,* or human transforming growth factor alpha. *Cancer Research* 56:4413–4423.

Greenlee, R. T., Murray, T., Bolden, S., and Wingo, P. A. 2000. Cancer statistics, 2000. *CA: A Cancer Journal for Clinicians* 50:7–33.

Grossman, L., Matanoski, G., Farmer, E., Hedayati, M., Ray, S., Trock, B., Hanfelt, J., Roush, G., Berwick, M., and Hu, J. J. 1999. DNA repair as a susceptibility factor in chronic diseases in human populations. In Dizdaroglu, M., and Karakaya, A. E., eds., *Advances in DNA Damage and Repair*, pp. 149–167. Kluwer Academic/Plenum Publishers, New York.

Gu, J., Zhao, H., Dinney, C. P., Zhu, Y., Leibovici, D., Bermejo, C. E., Grossman, H. B., and Wu, X. 2005. Nucleotide excision repair gene polymorphisms and recurrence after treatment for superficial bladder cancer. *Clinical Cancer Research* 11:1408–1415.

Guerrette, S., Acharya, S., and Fishel, R. 1999. The interaction of the human *MutL* homologues in hereditary nonpolyposis colon cancer. *Journal of Biological Chemistry* 274:6336–6341.

Guerrette, S., Wilson, T., Gradia, S., and Fishel, R. 1998. Interactions of human *hMSH2* with *hMSH3* and *hMSH2* with *hMSH6*: examination of mutations found in hereditary nonpolyposis colorectal cancer. *Molecular and Cellular Biology* 18:6616–6623.

Gutmann, D. H., and Collins, F. S. 2002. Neurofibromatosis I. In Vogelstein, B., and Kinzler, K. W., eds., *The Genetic Basis of Human Cancer* (2nd edition)., pp. 417–437. McGraw-Hill, New York.

Haenszel, W., and Kurihara, M. 1968. Studies of Japanese migrants. I. Mortality from cancer and other diseases among Japanese in the United States. *Journal of the National Cancer Institute* 40:43–68.

Hall, B. G. 2004. *Phylogenetic Trees Made Easy: A How-To Manual* (2nd edition). Sinauer Associates, Sunderland, MA.

Halpern, M. T., Gillespie, B. W., and Warner, K. E. 1993. Patterns of absolute risk of lung cancer mortality in former smokers. *Journal of the National Cancer Institute* 85:457–464.

Han, J., Colditz, G. A., Samson, L. D., and Hunter, D. J. 2004. Polymorphisms in DNA double-strand break repair genes and skin cancer risk. *Cancer Research* 64:3009–3013.

Hanahan, D., and Weinberg, R. A. 2000. The hallmarks of cancer. *Cell* 100:57–70.

Happle, R. 1993. Mosaicism in human skin. Understanding the patterns and mechanisms. *Archives of Dermatology* 129:1460–1470.

Happle, R. 1999. Loss of heterozygosity in human skin. *Journal of the American Academy of Dermatology* 41:143–164.

Happle, R., and Konig, A. 1999. Type 2 segmental manifestation of multiple glomus tumors: a review and reclassification of 5 case reports. *Dermatology* 198:270–272.

Harpending, H., and Cochran, G. 2006. Genetic diversity and genetic burden in humans. *Infection, Genetics and Evolution* 6:154–162.

Hartman, M., Czene, K., Reilly, M., Bergh, J., Lagiou, P., Trichopoulos, D., Adami, H. O., and Hall, P. 2005. Genetic implications of bilateral breast cancer: a population based cohort study. *Lancet Oncology* 6:377–382.

Hendry, J. H., Potten, C. S., Chadwick, C., and Bianchi, M. 1982. Cell death (apoptosis) in the mouse small intestine after low doses: effects of dose-rate, 14.7 MeV neutrons, and 600 MeV (maximum energy) neutrons. *International Journal of Radiation Biology and Related Studies in Physics, Chemistry and Medicine* 42:611–620.

Hethcote, H. W., and Knudson, A. G. 1978. Model for the incidence of embryonal cancers: application to retinoblastoma. *Proceedings of the National Academy of Sciences of the United States of America* 75:2453–2457.

Horiuchi, S., and Wilmoth, J. R. 1997. Age patterns of the life table aging rate for major causes of death in Japan, 1951-1990. *Journals of Gerontology. Series A, Biological Sciences and Medical Sciences* 52:B67–77.

Horiuchi, S., and Wilmoth, J. R. 1998. Deceleration in the age pattern of mortality at older ages. *Demography* 35:391–412.

Hotary, K. B., Allen, E. D., Brooks, P. C., Datta, N. S., Long, M. W., and Weiss, S. J. 2003. Membrane type I matrix metalloproteinase usurps tumor growth control imposed by the three-dimensional extracellular matrix. *Cell* 114:33–45.

Houle, D. 1992. Comparing evolvability and variability of quantitative traits. *Genetics* 130:195–204.

Hsieh, P. 2001. Molecular mechanisms of DNA mismatch repair. *Mutation Research* 486:71–87.

Hu, M., Yao, J., Cai, L., Bachman, K. E., van den Brule, F., Velculescu, V., and Polyak, K. 2005. Distinct epigenetic changes in the stromal cells of breast cancers. *Nature Genetics* 37:899–905.

Hunter, J. A. A., Savin, J., and Dahl, M. V. 1995. *Clinical Dermatology* (2nd edition). Blackwell Science, Oxford.

Hunter, K. D., Parkinson, E. K., and Harrison, P. R. 2005. Profiling early head and neck cancer. *Nature Reviews Cancer* 5:127–135.

Huntly, B. J., and Gilliland, D. G. 2005. Leukaemia stem cells and the evolution of cancer-stem-cell research. *Nature Reviews Cancer* 5:311–321.

Huson, S. M., Compston, D. A., Clark, P., and Harper, P. S. 1989. A genetic study of von Recklinghausen neurofibromatosis in south east Wales. I. Prevalence, fitness, mutation rate, and effect of parental transmission on severity. *Journal of Medical Genetics* 26:704–711.

Issa, J. P. 2000. CpG-island methylation in aging and cancer. *Current Topics in Microbiology and Immunology* 249:101–118.

Issa, J. P. 2004. Opinion: CpG island methylator phenotype in cancer. *Nature Reviews Cancer* 4:988–993.

Iversen, O. H. 1995. Of mice and men: a critical reappraisal of the two-stage theory of carcinogenesis. *Critical Reviews in Oncogenesis* 6:357–405.

Janes, S. M., Lowell, S., and Hutter, C. 2002. Epidermal stem cells. *Journal of Pathology* 197:479–491.

Jass, J. R. 2003. Serrated adenoma of the colorectum: a lesion with teeth. *American Journal of Pathology* 162:705–708.

Jass, J. R., Whitehall, V. L., Young, J., and Leggett, B. A. 2002a. Emerging concepts in colorectal neoplasia. *Gastroenterology* 123:862–876.

Jass, J. R., Young, J., and Leggett, B. A. 2002b. Evolution of colorectal cancer: change of pace and change of direction. *Journal of Gastroenterology and Hepatology* 17:17–26.

Jiang, W., Ananthaswamy, H. N., Muller, H. K., and Kripke, M. L. 1999. p53 protects against skin cancer induction by UV-B radiation. *Oncogene* 18:4247–4253.

Jonason, A. S., Kunala, S., Price, G. J., Restifo, R. J., Spinelli, H. M., Persing, J. A., Leffell, D. J., Tarone, R. E., and Brash, D. E. 1996. Frequent clones of *p53*-mutated keratinocytes in normal human skin. *Proceedings of the National Academy of Sciences of the United States of America* 93:14025–14029.

Jones, P. A., and Baylin, S. B. 2002. The fundamental role of epigenetic events in cancer. *Nature Reviews Genetics* 3:415–428.

Karpowicz, P., Morshead, C., Kam, A., Jervis, E., Ramunas, J., Cheng, V., and van der Kooy, D. 2005. Support for the immortal strand hypothesis: neural stem cells partition DNA asymmetrically in vitro. *Journal of Cell Biology* 170:721–732.

Kastan, M. B., and Bartek, J. 2004. Cell-cycle checkpoints and cancer. *Nature* 432:316–323.

Kemp, C. J., Donehower, L. A., Bradley, A., and Balmain, A. 1993. Reduction of *p53* gene dosage does not increase initiation or promotion but enhances malignant progression of chemically induced skin tumors. *Cell* 74:813–822.

Kim, B. G., Li, C., Qiao, W., Mamura, M., Kasperczak, B., Anver, M., Wolfraim, L., Hong, S., Mushinski, E., Potter, M., Kim, S. J., Fu, X. Y., Deng, C., and Letterio, J. J. 2006. Smad4 signalling in T cells is required for suppression of gastrointestinal cancer. *Nature* 441:1015–1019.

Kim, J. Y., Tavare, S., and Shibata, D. 2005. Counting human somatic cell replications: methylation mirrors endometrial stem cell divisions. *Proceedings of the National Academy of Sciences of the United States of America* 102:17739–17744.

Kim, J. Y., Tavare, S., and Shibata, D. 2006. Human hair genealogies and stem cell latency. *BMC Biology* 4:2.

Kim, K. M., Calabrese, P., Tavare, S., and Shibata, D. 2004. Enhanced stem cell survival in familial adenomatous polyposis. *American Journal of Pathology* 164:1369–1377.

Kim, K. M., and Shibata, D. 2002. Methylation reveals a niche: stem cell succession in human colon crypts. *Oncogene* 21:5441–5449.

Kim, K. M., and Shibata, D. 2004. Tracing ancestry with methylation patterns: most crypts appear distantly related in normal adult human colon. *BMC Gastroenterology* 4:8.

Kinzler, K. W., and Vogelstein, B. 1996. Lessons from hereditary colorectal cancer. *Cell* 87:159–170.

Kinzler, K. W., and Vogelstein, B. 1998. Landscaping the cancer terrain. *Science* 280:1036–1037.

Kinzler, K. W., and Vogelstein, B. 2002. Colorectal tumors. In Vogelstein, B., and Kinzler, K. W., eds., *The Genetic Basis of Human Cancer* (2nd edition)., pp. 583–612. McGraw-Hill, New York.

Kirkwood, T. B. 2005. Understanding the odd science of aging. *Cell* 120:437–447.

Klein, G. 1998. Foulds' dangerous idea revisited: the multistep development of tumors 40 years later. *Advances in Cancer Research* 72:1–23.

Knudson, A. G. 1971. Mutation and cancer: statistical study of retinoblastoma. *Proceedings of the National Academy of Sciences of the United States of America* 68:820–823.

Knudson, A. G. 1977. Genetic predisposition to cancer. In Hiatt, H. H., Watson, J. D., and Winsten, J. A., eds., *Origins of Human Cancer*, pp. 45–52. Cold Spring Harbor Publications, New York.

Knudson, A. G. 1993. Antioncogenes and human cancer. *Proceedings of the National Academy of Sciences of the United States of America* 90:10914–10921.

Knudson, A. G. 2001. Two genetic hits (more or less) to cancer. *Nature Reviews Cancer* 1:157–162.

Knudson, A. G. 2003. Cancer genetics through a personal retrospectroscope. *Genes, Chromosomes and Cancer* 38:288–291.

Knudson, A. G., Hethcote, H. W., and Brown, B. W. 1975. Mutation and childhood cancer: a probabilistic model for the incidence of retinoblastoma. *Proceedings of the National Academy of Sciences of the United States of America* 72:5116–5120.

Kohler, S. W., Provost, G. S., Fieck, A., Kretz, P. L., Bullock, W. O., Sorge, J. A., Putman, D. L., and Short, J. M. 1991. Spectra of spontaneous and mutagen-induced mutations in the lacI gene in transgenic mice. *Proceedings of the National Academy of Sciences of the United States of America* 88:7958–7962.

Komarova, N. L., Sengupta, A., and Nowak, M. A. 2003. Mutation-selection networks of cancer initiation: tumor suppressor genes and chromosomal instability. *Journal of Theoretical Biology* 223:433–450.

Kondo, M., Wagers, A. J., Manz, M. G., Prohaska, S. S., Scherer, D. C., Beilhack, G. F., Shizuru, J. A., and Weissman, I. L. 2003. Biology of hematopoietic stem cells and progenitors: implications for clinical application. *Annual Review of Immunology* 21:759–806.

Kroemer, G. 2004. Cell death and cancer: an introduction. *Oncogene* 23:2744–2745.

Kuo, M. H., and Allis, C. D. 1998. Roles of histone acetyltransferases and deacetylases in gene regulation. *Bioessays* 20:615–626.

Kwabi-Addo, B., Giri, D., Schmidt, K., Podsypanina, K., Parsons, R., Greenberg, N., and Ittmann, M. 2001. Haploinsufficiency of the *Pten* tumor suppressor gene promotes prostate cancer progression. *Proceedings of the National Academy of Sciences of the United States of America* 98:11563–11568.

Lajtha, L. G. 1979. Stem cell concepts. *Differentiation* 14:23–34.

Lamlum, H., Al Tassan, N., Jaeger, E., Frayling, I. M., Sieber, O., Reza, F. B., Eckert, M., Rowan, A., Barclay, E., Atkin, W., Williams, C. B., Gilbert, J., Cheadle, J., Bell, J. A., Houlston, R., Bodmer, W. F., Sampson, J., and Tomlinson, I. P. 2000. Germline *APC* variants in patients with multiple colorectal adenomas, with evidence for the particular importance of E1317Q. *Human Molecular Genetics* 9:2215–2221.

Lang, D., Lu, M. M., Huang, L., Engleka, K. A., Zhang, M., Chu, E. Y., Lipner, S., Skoultchi, A., Millar, S. E., and Epstein, J. A. 2005. Pax3 functions at a nodal point in melanocyte stem cell differentiation. *Nature* 433:884–887.

Lawley, P. D. 1994. Historical origins of current concepts of carcinogenesis. *Advances in Cancer Research* 65:17–111.

Lechler, T., and Fuchs, E. 2005. Asymmetric cell divisions promote stratification and differentiation of mammalian skin. *Nature* 437:275–280.

Lee, C. 2002. Irresistible force meets immovable object: SNP mapping of complex diseases. *Trends in Genetics* 18:67–69.

Lichten, M., and Haber, J. E. 1989. Position effects in ectopic and allelic mitotic recombination in *Saccharomyces cerevisiae*. *Genetics* 123:261–268.

Lichtenstein, P., Holm, N. V., Verkasalo, P. K., Iliadou, A., Kaprio, J., Koskenvuo, M., Pukkala, E., Skytthe, A., and Hemminki, K. 2000. Environmental and heritable factors in the causation of cancer–analyses of cohorts of twins from Sweden, Denmark, and Finland. *New England Journal of Medicine* 343:78–85.

Limpert, E., Stahel, W. A., and Abbt, M. 2001. Log-normal distributions across the sciences: keys and clues. *Bioscience* 51:341–352.

Liotta, L. A., and Kohn, E. C. 2001. The microenvironment of the tumour-host interface. *Nature* 411:375.

Lipkin, S. M., Wang, V., Jacoby, R., Banerjee-Basu, S., Baxevanis, A. D., Lynch,

H. T., Elliott, R. M., and Collins, F. S. 2000. *MLH3:* a DNA mismatch repair gene associated with mammalian microsatellite instability. *Nature Genetics* 24:27–35.

Liu, S., Dontu, G., and Wicha, M. S. 2005. Mammary stem cells, self-renewal pathways, and carcinogenesis. *Breast Cancer Research* 7:86–95.

Loeb, L. A. 1991. Mutator phenotype may be required for multistage carcinogenesis. *Cancer Research* 51:3075–3079.

Loeb, L. A. 1998. Cancer cells exhibit a mutator phenotype. *Advances in Cancer Research* 72:25–56.

Loeb, L. A., Springgate, C. F., and Battula, N. 1974. Errors in DNA replication as a basis of malignant changes. *Cancer Research* 34:2311–2321.

Lowe, S. W., Cepero, E., and Evan, G. 2004. Intrinsic tumour suppression. *Nature* 432:307–315.

Luebeck, E. G., and Moolgavkar, S. H. 2002. Multistage carcinogenesis and the incidence of colorectal cancer. *Proceedings of the National Academy of Sciences of the United States of America* 99:15095–15100.

Luria, S. E., and Delbrück, M. 1943. Mutations of bacteria from virus sensitivity to virus resistance. *Genetics* 28:491–511.

Lutz, W. K. 1999. Dose-response relationships in chemical carcinogenesis reflect differences in individual susceptibility. Consequences for cancer risk assessment, extrapolation, and prevention. *Human and Experimental Toxicology* 18:707–712.

Lynch, H. T., Smyrk, T., and Jass, J. R. 1995. Hereditary nonpolyposis colorectal cancer and colonic adenomas: aggressive adenomas? *Seminars in Surgical Oncology* 11:406–410.

Lynch, M., and Walsh, B. 1998. *Genetics and Analysis of Quantitative Traits.* Sinauer, Sunderland, MA.

MacKenzie, I., and Rous, P. 1940. The experimental disclosure of latent neoplastic transformation in tarred skin. *Journal of Experimental Medicine* 71:391–416.

Maley, C. C., Galipeau, P. C., Finley, J. C., Wongsurawat, V. J., Li, X., Sanchez, C. A., Paulson, T. G., Blount, P. L., Risques, R. A., Rabinovitch, P. S., and Reid, B. J. 2006. Genetic clonal diversity predicts progression to esophageal adenocarcinoma. *Nature Genetics* 38:468–473.

Mao, J. H., Lindsay, K. A., Balmain, A., and Wheldon, T. E. 1998. Stochastic modelling of tumorigenesis in *p53* deficient mice. *British Journal of Cancer* 77:243–252.

Marsh, D., and Zori, R. 2002. Genetic insights into familial cancers—update and recent discoveries. *Cancer Letters* 181:125–164.

Marshman, E., Booth, C., and Potten, C. S. 2002. The intestinal epithelial stem cell. *Bioessays* 24:91–98.

Mathon, N. F., and Lloyd, A. C. 2001. Cell senescence and cancer. *Nature Reviews Cancer* 1:203–213.

Maynard Smith, J., and Szathmary, E. 1995. *The Major Transitions in Evolution.* W. H. Freeman, New York.

Merok, J. R., Lansita, J. A., Tunstead, J. R., and Sherley, J. L. 2002. Cosegregation of chromosomes containing immortal DNA strands in cells that cycle with asymmetric stem cell kinetics. *Cancer Research* 62:6791-6795.

Meza, R., Luebeck, E. G., and Moolgavkar, S. H. 2005. Gestational mutations and carcinogenesis. *Mathematical Biosciences* 197:188-210.

Michor, F., Iwasa, Y., Komarova, N. L., and Nowak, M. A. 2003. Local regulation of homeostasis favors chromosomal instability. *Current Biology* 13:581-584.

Michor, F., Iwasa, Y., and Nowak, M. A. 2004. Dynamics of cancer progression. *Nature Reviews Cancer* 4:197-205.

Mitchell, R. J., Farrington, S. M., Dunlop, M. G., and Campbell, H. 2002. Mismatch repair genes *hMLH1* and *hMSH2* and colorectal cancer: a HuGE review. *American Journal of Epidemiology* 156:885-902.

Mohrenweiser, H. W., Wilson, D. M., and Jones, I. M. 2003. Challenges and complexities in estimating both the functional impact and the disease risk associated with the extensive genetic variation in human DNA repair genes. *Mutation Research* 526:93-125.

Moolgavkar, S. H. 1978. The multistage theory of carcinogenesis and the age distribution of cancer in man. *Journal of the National Cancer Institute* 61:49-52.

Moolgavkar, S. H. 2004. Commentary: Fifty years of the multistage model: remarks on a landmark paper. *International Journal of Epidemiology* 33:1182-1183.

Moolgavkar, S. H., Dewanji, A., and Luebeck, E. G. 1989. Cigarette smoking and lung cancer: reanalysis of the British doctors' data. *Journal of the National Cancer Institute* 81:415-420.

Moolgavkar, S. H., and Knudson, A. G. 1981. Mutation and cancer: a model for human carcinogenesis. *Journal of the National Cancer Institute* 66:1037-1052.

Moolgavkar, S. H., Krewski, D., and Schwarz, M. 1999. Mechanisms of carcinogenesis and biologically-based models for quantitative estimation and prediction of cancer risk. In Moolgavkar, S. H., Krewski, D., Zeise, L., Cardis, E., and Moller, H., eds., *Quantitative Estimation and Prediction of Cancer Risk,* pp. 179-238. IARC Scientific Publications, Lyon.

Moolgavkar, S. H., and Venzon, D. J. 1979. Two-event models for carcinogenesis: incidence curves for childhood and adult tumors. *Mathematical Biosciences* 47:55-77.

Morris, R. J. 2004. A perspective on keratinocyte stem cells as targets for skin carcinogenesis. *Differentiation* 72:381-386.

Morris, R. J., Liu, Y., Marles, L., Yang, Z., Trempus, C., Li, S., Lin, J. S., Sawicki, J. A., and Cotsarelis, G. 2004. Capturing and profiling adult hair follicle stem cells. *Nature Biotechnology* 22:411-417.

Morrison, S. J., and Kimble, J. 2006. Asymmetric and symmetric stem-cell divisions in development and cancer. *Nature* 441:1068–1074.

Mousseau, T. A., and Roff, D. A. 1987. Natural selection and the heritability of fitness components. *Heredity* 59:181–197.

Mueller, M. M., and Fusenig, N. E. 2004. Friends or foes—bipolar effects of the tumour stroma in cancer. *Nature Reviews Cancer* 4:839–849.

Muller, H. J. 1951. Radiation damage to the genetic material. In Baitsell, G. A., ed., *Science in Progress: Seventh Series*, pp. 93–165. Yale University Press, New Haven.

Murray, J. D. 1989. *Mathematical Biology.* Springer-Verlag, New York.

Newsham, I. F., Hadjistilianou, T., and Cavenee, W. K. 2002. Retinoblastoma. In Vogelstein, B., and Kinzler, K. W., eds., *The Genetic Basis of Human Cancer* (2nd edition)., pp. 357–386. McGraw-Hill, New York.

Ng, P. C., and Henikoff, S. 2003. SIFT: predicting amino acid changes that affect protein function. *Nucleic Acids Research* 31:3812–3814.

Nishimura, E. K., Jordan, S. A., Oshima, H., Yoshida, H., Osawa, M., Moriyama, M., Jackson, I. J., Barrandon, Y., Miyachi, Y., and Nishikawa, S. 2002. Dominant role of the niche in melanocyte stem-cell fate determination. *Nature* 416:854–860.

Nordling, C. O. 1953. A new theory on cancer-inducing mechanism. *British Journal of Cancer* 7:68–72.

Nowak, M. A., Komarova, N. L., Sengupta, A., Jallepalli, P. V., Shih, I., Vogelstein, B., and Lengauer, C. 2002. The role of chromosomal instability in tumor initiation. *Proceedings of the National Academy of Sciences of the United States of America* 99:16226–16231.

Nowell, P. C. 1976. The clonal evolution of tumor cell populations. *Science* 194:23–28.

Nunney, L. 1999. Lineage selection and the evolution of multistage carcinogenesis. *Proceedings of the Royal Society of London. Series B: Biological Sciences* 266:493–498.

Nunney, L. 2003. The population genetics of multistage carcinogenesis. *Proceedings of the Royal Society of London. Series B: Biological Sciences* 270:1183–1191.

Page, R. D. M., and Holmes, E. C. 1998. *Molecular Evolution: A Phylogenetic Approach.* Blackwell Scientific, Oxford.

Pardal, R., Clarke, M. F., and Morrison, S. J. 2003. Applying the principles of stem-cell biology to cancer. *Nature Reviews Cancer* 3:895–902.

Park, M. 2002. Oncogenes. In Vogelstein, B., and Kinzler, K. W., eds., *The Genetic Basis of Human Cancer* (2nd edition)., pp. 177–196. McGraw-Hill, New York.

Park, S. J., Rashid, A., Lee, J. H., Kim, S. G., Hamilton, S. R., and Wu, T. T. 2003. Frequent CpG island methylation in serrated adenomas of the colorectum. *American Journal of Pathology* 162:815–822.

Parkin, D. M., Whelan, S. L., Ferlay, J., Teppo, L., and Thomas, D. B., eds. 2002.

Cancer Incidence in Five Continents, Vol. VIII. International Agency for Research on Cancer, Lyon, France.

Pelengaris, S., Khan, M., and Evan, G. 2002. *c-MYC:* more than just a matter of life and death. *Nature Reviews Cancer* 2:764–776.

Peltomaki, P., and Vasen, H. F. 1997. Mutations predisposing to hereditary nonpolyposis colorectal cancer: database and results of a collaborative study. The International Collaborative Group on Hereditary Nonpolyposis Colorectal Cancer. *Gastroenterology* 113:1146–1158.

Peto, J. 2001. Cancer epidemiology in the last century and the next decade. *Nature* 411:390–395.

Peto, J., Collins, N., Barfoot, R., Seal, S., Warren, W., Rahman, N., Easton, D. F., Evans, C., Deacon, J., and Stratton, M. R. 1999. Prevalence of *BRCA1* and *BRCA2* gene mutations in patients with early-onset breast cancer. *Journal of the National Cancer Institute* 91:943–949.

Peto, J., and Mack, T. M. 2000. High constant incidence in twins and other relatives of women with breast cancer. *Nature Genetics* 26:411–414.

Peto, R. 1977. Epidemiology, multistage models and short-term mutagenicity tests. In Hiatt, H. H., Watson, J. D., and Winsten, J. A., eds., *Origins of Human Cancer*, pp. 1403–1428. Cold Spring Harbor Publications, New York.

Peto, R., Darby, S., Deo, H., Silcocks, P., Whitley, E., and Doll, R. 2000. Smoking, smoking cessation, and lung cancer in the UK since 1950: combination of national statistics with two case-control studies. *British Medical Journal* 321:323–329.

Peto, R., Gray, R., Brantom, P., and Grasso, P. 1991. Dose and time relationships for tumor induction in the liver and esophagus of 4080 inbred rats by chronic ingestion of N-nitrosodiethylamine or N-nitrosodimethylamine. *Cancer Research* 51:6452–6469.

Pfeifer, G. P., Steigerwald, S. D., Hansen, R. S., Gartler, S. M., and Riggs, A. D. 1990. Polymerase chain reaction-aided genomic sequencing of an X chromosome-linked CpG island: methylation patterns suggest clonal inheritance, CpG site autonomy, and an explanation of activity state stability. *Proceedings of the National Academy of Sciences of the United States of America* 87:8252–8256.

Pharoah, P. D., Antoniou, A., Bobrow, M., Zimmern, R. L., Easton, D. F., and Ponder, B. A. 2002. Polygenic susceptibility to breast cancer and implications for prevention. *Nature Genetics* 31:33–36.

Pierce, D. A., and Vaeth, M. 2003. Age-time patterns of cancer to be anticipated from exposure to general mutagens. *Biostatistics* 4:231–248.

Pike, M. C., Krailo, M. D., Henderson, B. E., Casagrande, J. T., and Hoel, D. G. 1983. "Hormonal" risk factors, "breast tissue age" and the age-incidence of breast cancer. *Nature* 303:767–770.

Pike, M. C., Pearce, C. L., and Wu, A. H. 2004. Prevention of cancers of the breast, endometrium and ovary. *Oncogene* 23:6379–6391.

Platt, R. 1955. Clonal ageing and cancer. *Lancet* 265:867.

Pletcher, S. D., and Curtsinger, J. W. 1998. Mortality plateaus and the evolution of senescence: why are old-age mortality rates so low? *Evolution* 52:454-464.

Popanda, O., Schattenberg, T., Phong, C. T., Butkiewicz, D., Risch, A., Edler, L., Kayser, K., Dienemann, H., Schulz, V., Drings, P., Bartsch, H., and Schmezer, P. 2004. Specific combinations of DNA repair gene variants and increased risk for non-small cell lung cancer. *Carcinogenesis* 25:2433-2441.

Potten, C. S. 1974. The epidermal proliferative unit: the possible role of the central basal cell. *Cell and Tissue Kinetics* 7:77-88.

Potten, C. S. 1977. Extreme sensitivity of some intestinal crypt cells to X and gamma irradiation. *Nature* 269:518-521.

Potten, C. S. 1981. Cell replacement in epidermis (keratopoiesis) via discrete units of proliferation. *International Review of Cytology* 69:271-318.

Potten, C. S. 1998. Stem cells in gastrointestinal epithelium: numbers, characteristics and death. *Philosophical Transactions of the Royal Society of London. Series B: Biological Sciences* 353:821-830.

Potten, C. S., and Booth, C. 2002. Keratinocyte stem cells: a commentary. *Journal of Investigative Dermatology* 119:888-899.

Potten, C. S., and Grant, H. K. 1998. The relationship between ionizing radiation-induced apoptosis and stem cells in the small and large intestine. *British Journal of Cancer* 78:993-1003.

Potten, C. S., Li, Y. Q., O'Connor, P. J., and Winton, D. J. 1992. A possible explanation for the differential cancer incidence in the intestine, based on distribution of the cytotoxic effects of carcinogens in the murine large bowel. *Carcinogenesis* 13:2305-2312.

Potten, C. S., Owen, G., and Booth, D. 2002. Intestinal stem cells protect their genome by selective segregation of template DNA strands. *Journal of Cell Science* 115:2381-2388.

Prehn, R. T. 2005. The role of mutation in the new cancer paradigm. *Cancer Cell International* 5:9.

Preston-Martin, S., Pike, M. C., Ross, R. K., Jones, P. A., and Henderson, B. E. 1990. Increased cell division as a cause of human cancer. *Cancer Research* 50:7415-7421.

R Development Core Team 2004. *R: A Language and Environment for Statistical Computing.* R Foundation for Statistical Computing, Vienna.

Rajagopalan, H., Bardelli, A., Lengauer, C., Kinzler, K. W., Vogelstein, B., and Velculescu, V. E. 2002. Tumorigenesis: *RAF/RAS* oncogenes and mismatch-repair status. *Nature* 418:934.

Rajagopalan, H., Nowak, M. A., Vogelstein, B., and Lengauer, C. 2003. The significance of unstable chromosomes in colorectal cancer. *Nature Reviews Cancer* 3:695-701.

Rambhatla, L., Ram-Mohan, S., Cheng, J. J., and Sherley, J. L. 2005. Immortal DNA strand cosegregation requires *p53/IMPDH*-dependent asymmetric self-renewal associated with adult stem cells. *Cancer Research* 65:3155-3161.

Ramensky, V., Bork, P., and Sunyaev, S. 2002. Human non-synonymous SNPs: server and survey. *Nucleic Acids Research* 30:3894-3900.

Reya, T., Morrison, S. J., Clarke, M. F., and Weissman, I. L. 2001. Stem cells, cancer, and cancer stem cells. *Nature* 414:105-111.

Ries, L. A. G., Smith, M. A., Gurney, J. G., Linet, M., Tamra, T., Young, J. L., and Bunin, G. R., eds. 1999. *Cancer Incidence and Survival among Children and Adolescents: United States SEER Program 1975-1995*, NIH Pub. No. 99-4649. National Cancer Institute, SEER Program, Bethesda, MD [http://seer.cancer.gov/publications/childhood/].

Rizzo, S., Attard, G., and Hudson, D. L. 2005. Prostate epithelial stem cells. *Cell Proliferation* 38:363-374.

Ro, S., and Rannala, B. 2001. Methylation patterns and mathematical models reveal dynamics of stem cell turnover in the human colon. *Proceedings of the National Academy of Sciences of the United States of America* 98:10519-10521.

Roberts, S. A., Spreadborough, A. R., Bulman, B., Barber, J. B., Evans, D. G., and Scott, D. 1999. Heritability of cellular radiosensitivity: a marker of low-penetrance predisposition genes in breast cancer? *American Journal of Human Genetics* 65:784-794.

Robertson, K. D. 2005. DNA methylation and human disease. *Nature Reviews Genetics* 6:597-610.

Rose, M. R. 1991. *Evolutionary Biology of Aging*. Oxford University Press, Oxford.

Rose, M. R., and Mueller, L. D. 2000. Ageing and immortality. *Philosophical Transactions of the Royal Society of London, Series B* 355:1657-1662.

Rous, P., and Kidd, I. G. 1941. Conditional neoplasma and subthreshold neoplastic states: a study of tar tumors in rabbits. *Journal of Experimental Medicine* 73:365-389.

Rubin, H., ed. 2005. *Microenvironmental Regulation of Tumor Development*, Vol. 15 of *Seminars in Cancer Biology*. Saunders Scientific Publications, Philadelphia.

Rudolph, R. 1993. Familial multiple glomangiomas. *Annals of Plastic Surgery* 30:183-185.

Samuelsson, B., and Axelsson, R. 1981. Neurofibromatosis. A clinical and genetic study of 96 cases in Gothenburg, Sweden. *Acta Dermato-Venereologica. Supplementum* 95:67-71.

Scherer, E., and Emmelot, P. 1979. Multihit kinetics of tumor cell formation and risk assessment of low doses of carcinogen. In Griffin, A. C., and Shaw, C. R., eds., *Carcinogens: Identification and Mechanisms of Action*, pp. 337-364. Raven Press, New York.

Seligson, D. B., Horvath, S., Shi, T., Yu, H., Tze, S., Grunstein, M., and Kurdistani, S. K. 2005. Global histone modification patterns predict risk of prostate cancer recurrence. *Nature* 435:1262-1266.

Sell, S. 2004. Stem cell origin of cancer and differentiation therapy. *Critical Reviews in Oncology/Hematology* 51:1-28.

Sergeyev, A. S. 1975. On the mutation rate of neurofibromatosis. *Humangenetik* 28:129-138.

Shibata, D. 2006. Clonal diversity in tumor progression. *Nature Genetics* 38:402-403.

Shibata, D., and Tavare, S. 2006. Counting divisions in a human somatic cell tree: how, what and why? *Cell Cycle* 5:610-614.

Shimodaira, H., Filosi, N., Shibata, H., Suzuki, T., Radice, P., Kanamaru, R., Friend, S. H., Kolodner, R. D., and Ishioka, C. 1998. Functional analysis of human *MLH1* mutations in *Saccharomyces cerevisiae*. *Nature Genetics* 19:384-389.

Shizuru, J. A., Negrin, R. S., and Weissman, I. L. 2005. Hematopoietic stem and progenitor cells: clinical and preclinical regeneration of the hematolymphoid system. *Annual Review of Medicine* 56:509-538.

Shmookler Reis, R. J., and Goldstein, S. 1982. Variability of DNA methylation patterns during serial passage of human diploid fibroblasts. *Proceedings of the National Academy of Sciences of the United States of America* 79:3949-3953.

Sieber, O., Heinimann, K., and Tomlinson, I. P. 2005. Genomic stability and tumorigenesis. *Seminars in Cancer Biology* 15:61-66.

Siegel, D. H., and Sybert, V. P. 2006. Mosaicism in genetic skin disorders. *Pediatric Dermatology* 23:87-92.

Simon, R., Eltze, E., Schafer, K. L., Burger, H., Semjonow, A., Hertle, L., Dockhorn-Dworniczak, B., Terpe, H. J., and Bocker, W. 2001. Cytogenetic analysis of multifocal bladder cancer supports a monoclonal origin and intraepithelial spread of tumor cells. *Cancer Research* 61:355-362.

Singh, S. K., Clarke, I. D., Hide, T., and Dirks, P. B. 2004. Cancer stem cells in nervous system tumors. *Oncogene* 23:7267-7273.

Slaga, T. J., Budunova, I. V., Gimenez-Conti, I. B., and Aldaz, C. M. 1996. The mouse skin carcinogenesis model. *Journal of Investigative Dermatology. Symposium Proceedings* 1:151-156.

Slaughter, D. P., Southwick, H. W., and Smejkal, W. 1953. Field cancerization in oral stratified squamous epithelium; clinical implications of multicentric origin. *Cancer* 6:963-968.

Smalley, K. S., Brafford, P. A., and Herlyn, M. 2005. Selective evolutionary pressure from the tissue microenvironment drives tumor progression. *Seminars in Cancer Biology* 15:451-459.

Smith, G. H. 2005. Label-retaining epithelial cells in mouse mammary gland divide asymmetrically and retain their template DNA strands. *Development* 132:681-687.

Sommer, L. 2005. Checkpoints of melanocyte stem cell development. *Science's STKE* 2005:pe42.

Sontag, L. B., Lorincz, M. C., and Luebeck, E. G. 2006. Dynamics, stability and inheritance of somatic DNA methylation imprints. *Journal of Theoretical Biology* 242:890-899.

Stein, W. D. 1991. Analysis of cancer incidence data on the basis of multistage and clonal growth models. *Advances in Cancer Research* 56:161-213.

Stellman, S. D., Boffetta, P., and Garfinkel, L. 1988. Smoking habits of 800,000 American men and women in relation to their occupations. *American Journal of Industrial Medicine* 13:43-58.

Stocks, P. 1953. A study of the age curve for cancer of the stomach in connection with a theory of the cancer producing mechanism. *British Journal of Cancer* 7:407-417.

Storm, S. M., and Rapp, U. R. 1993. Oncogene activation: *c-raf-1* gene mutations in experimental and naturally occurring tumors. *Toxicology Letters* 67:201-210.

Struewing, J. P., Hartge, P., Wacholder, S., Baker, S. M., Berlin, M., McAdams, M., Timmerman, M. M., Brody, L. C., and Tucker, M. A. 1997. The risk of cancer associated with specific mutations of *BRCA1* and *BRCA2* among Ashkenazi Jews. *New England Journal of Medicine* 336:1401-1408.

Taibjee, S. M., Bennett, D. C., and Moss, C. 2004. Abnormal pigmentation in hypomelanosis of Ito and pigmentary mosaicism: the role of pigmentary genes. *British Journal of Dermatology* 151:269-282.

Takano, H., Ema, H., Sudo, K., and Nakauchi, H. 2004. Asymmetric division and lineage commitment at the level of hematopoietic stem cells: inference from differentiation in daughter cell and granddaughter cell pairs. *Journal of Experimental Medicine* 199:295-302.

Tan, W. Y. 1991. *Stochastic Models of Carcinogenesis*. Marcel Dekker, New York.

Taniguchi, K., Roberts, L. R., Aderca, I. N., Dong, X., Qian, C., Murphy, L. M., Nagorney, D. M., Burgart, L. J., Roche, P. C., Smith, D. I., Ross, J. A., and Liu, W. 2002. Mutational spectrum of *beta-catenin, AXIN1*, and *AXIN2* in hepatocellular carcinomas and hepatoblastomas. *Oncogene* 21:4863-4871.

Tannergard, P., Lipford, J. R., Kolodner, R., Frodin, J. E., Nordenskjold, M., and Lindblom, A. 1995. Mutation screening in the *hMLH1* gene in Swedish hereditary nonpolyposis colon cancer families. *Cancer Research* 55:6092-6096.

Taylor, G., Lehrer, M. S., Jensen, P. J., Sun, T. T., and Lavker, R. M. 2000. Involvement of follicular stem cells in forming not only the follicle but also the epidermis. *Cell* 102:451-461.

Thompson, L. H., and Schild, D. 2002. Recombinational DNA repair and human disease. *Mutation Research* 509:49-78.

Tomlinson, I. P., Novelli, M. R., and Bodmer, W. F. 1996. The mutation rate and cancer. *Proceedings of the National Academy of Sciences of the United States of America* 93:14800-14803.

Tonin, P., Serova, O., Lenoir, G., Lynch, H. T., Durocher, F., Simard, J., Morgan, K., and Narod, S. 1995. *BRCA1* mutations in Ashkenazi Jewish women. *American Journal of Human Genetics* 57:189.

Tsao, J. L., Tavare, S., Salovaara, R., Jass, J. R., Aaltonen, L. A., and Shibata, D. 1999. Colorectal adenoma and cancer divergence. Evidence of multilineage progression. *American Journal of Pathology* 154:1815-1824.

Turker, M. S. 2003. Autosomal mutation in somatic cells of the mouse. *Mutagenesis* 18:1-6.

Twort, C. C., and Twort, J. M. 1928. Observations on the reaction of the skin to oils and tars. *Journal of Hygiene* 28:219-227.

Twort, J. M., and Twort, C. C. 1939. Comparative activity of some carcinogenic hydrocarbons. *American Journal of Cancer* 35:80-85.

Tyzzer, E. E. 1916. Tumor immunity. *Journal of Cancer Research* 1:125-156.

van Kempen, L. C., Ruiter, D. J., van Muijen, G. N., and Coussens, L. M. 2003. The tumor microenvironment: a critical determinant of neoplastic evolution. *European Journal of Cell Biology* 82:539-548.

Vaupel, J. W. 2003. Post-darwinian longevity. In Carey, J. R., and Tuljapurkar, S., eds., *Life Span: Evolutionary, Ecological, and Demographic Perspectives*, pp. 258-269. Population Council, New York.

Vaupel, J. W., Carey, J. R., Christensen, K., Johnson, T. E., Yashin, A. I., Holm, N. V., Iachine, I. A., Kannisto, V., Khazaeli, A. A., Liedo, P., Longo, V. D., Zeng, Y., Manton, K. G., and Curtsinger, J. W. 1998. Biodemographic trajectories of longevity. *Science* 280:855-860.

Veale, A. M. O. 1965. *Intestinal Polyposis*, Vol. 40 of *Eugenics Laboratory Memoirs*. Cambridge University Press, Cambridge.

Vijg, J., and Dolle, M. E. 2002. Large genome rearrangements as a primary cause of aging. *Mechanisms of Ageing and Development* 123:907-915.

Vikkula, M., Boon, L. M., and Mulliken, J. B. 2001. Molecular genetics of vascular malformations. *Matrix Biology* 20:327-335.

Villadsen, R. 2005. In search of a stem cell hierarchy in the human breast and its relevance to breast cancer evolution. *APMIS: Acta Pathologica, Microbiologica, et Immunologica Scandinavica* 113:903-921.

Vineis, P., Alavanja, M., Buffler, P., Fontham, E., Franceschi, S., Gao, Y. T., Gupta, P. C., Hackshaw, A., Matos, E., Samet, J., Sitas, F., Smith, J., Stayner, L., Straif, K., Thun, M. J., Wichmann, H. E., Wu, A. H., Zaridze, D., Peto, R., and Doll, R. 2004. Tobacco and cancer: recent epidemiological evidence. *Journal of the National Cancer Institute* 96:99-106.

Vineis, P., and Pirastu, R. 1997. Aromatic amines and cancer. *Cancer Causes and Control* 8:346-355.

Vogelstein, B., and Kinzler, K. W., eds. 2002. *The Genetic Basis of Human Cancer* (2nd edition). McGraw-Hill, New York.

Wallace, D. C. 2005. A mitochondrial paradigm of metabolic and degenerative diseases, aging, and cancer: a dawn for evolutionary medicine. *Annual Review of Genetics* 39:359-407.

Wang, C., Fu, M., Mani, S., Wadler, S., Senderowicz, A. M., and Pestell, R. G. 2001. Histone acetylation and the cell-cycle in cancer. *Frontiers in Bioscience* 6:D610-629.

Watt, F. M. 1998. Epidermal stem cells: markers, patterning and the control of stem cell fate. *Philosophical Transactions of the Royal Society of London. Series B: Biological Sciences* 353:831–837.

Watt, F. M., and Hogan, B. L. 2000. Out of Eden: stem cells and their niches. *Science* 287:1427–1430.

Webster, M. T., Rozycka, M., Sara, E., Davis, E., Smalley, M., Young, N., Dale, T. C., and Wooster, R. 2000. Sequence variants of the axin gene in breast, colon, and other cancers: an analysis of mutations that interfere with *GSK3* binding. *Genes, Chromosomes and Cancer* 28:443–453.

Weinberg, R. A. 2007. *The Biology of Cancer.* Garland Science, New York.

Weiss, K. M., and Terwilliger, J. D. 2000. How many diseases does it take to map a gene with SNPs? *Nature Genetics* 26:151–157.

Weiss, L. 2000. Heterogeneity of cancer cell populations and metastasis. *Cancer and Metastasis Reviews* 19:351–379.

Weissman, I. L. 2000. Stem cells: units of development, units of regeneration, and units in evolution. *Cell* 100:157–168.

Welcsh, P. L., and King, M. C. 2001. *BRCA1* and *BRCA2* and the genetics of breast and ovarian cancer. *Human Molecular Genetics* 10:705–713.

Whittemore, A. S. 1977. The age distribution of human cancer for carcinogenic exposures of varying intensity. *American Journal of Epidemiology* 106:418–432.

Whittemore, A. S. 1988. Effect of cigarette smoking in epidemiological studies of lung cancer. *Statistics in Medicine* 7:223–238.

Whittemore, A. S., and Keller, J. B. 1978. Quantitative theories of carcinogenesis. *SIAM Review* 20:1–30.

Wijnen, J., Khan, P. M., Vasen, H. F., Menko, F., van der Klift, H., van den Broek, M., van Leeuwen-Cornelisse, I., Nagengast, F., Meijers-Heijboer, E. J., Lindhout, D., Griffioen, G., Cats, A., Kleibeuker, J., Varesco, L., Bertario, L., Bisgaard, M. L., Mohr, J., Kolodner, R., and Fodde, R. 1996. Majority of *hMLH1* mutations responsible for hereditary nonpolyposis colorectal cancer cluster at the exonic region 15-16. *American Journal of Human Genetics* 58:300–307.

Witkowski, J. A. 1990. The inherited character of cancer—an historical survey. *Cancer Cells* 2:228–257.

Wright, A., Charlesworth, B., Rudan, I., Carothers, A., and Campbell, H. 2003. A polygenic basis for late-onset disease. *Trends in Genetics* 19:97–106.

Wu, X., Gu, J., Grossman, H. B., Amos, C. I., Etzel, C., Huang, M., Zhang, Q., Millikan, R. E., Lerner, S., Dinney, C. P., and Spitz, M. R. 2006. Bladder cancer predisposition: a multigenic approach to DNA-repair and cell-cycle-control genes. *American Journal of Human Genetics* 78:464–479.

Wunderlich, V. 2002. Chromosomes and cancer: Theodor Boveri's predictions 100 years later. *Journal of Molecular Medicine* 80:545–548.

Yamada, K. M. 2003. Cell biology: tumour jailbreak. *Nature* 424:889–890.

Yamashita, Y. M., Jones, D. L., and Fuller, M. T. 2003. Orientation of asymmetric stem cell division by the APC tumor suppressor and centrosome. *Science* 301:1547–1550.

Yatabe, Y., Tavare, S., and Shibata, D. 2001. Investigating stem cells in human colon by using methylation patterns. *Proceedings of the National Academy of Sciences of the United States of America* 98:10839–10844.

Young, J. L., Smith, M. A., Roffers, S. D., Liff, J. M., and Bunin, G. R. 1999. Retinoblastoma. In Ries, L. A. G., Smith, M. A., Gurney, J. G., Linet, M., Tamra, T., Young, J. L., and Bunin, G. R., eds., *Cancer Incidence and Survival among Children and Adolescents: United States SEER Program 1975-1995*, NIH Pub. No. 99-4649, pp. 73–78. National Cancer Institute, SEER Program, Bethesda, MD [http://seer.cancer.gov/publications/childhood/].

Yuan, L. W., and Keil, R. L. 1990. Distance-independence of mitotic intrachromosomal recombination in *Saccharomyces cerevisiae. Genetics* 124:263–273.

Zaghloul, N. A., Yan, B., and Moody, S. A. 2005. Step-wise specification of retinal stem cells during normal embryogenesis. *Biologie Cellulaire* 97:321–337.

Zeise, L., Wilson, R., and Crouch, E. A. 1987. Dose-response relationships for carcinogens: a review. *Environmental Health Perspectives* 73:259–306.

Zheng, Q. 1999. Progress of a half century in the study of the Luria-Delbruck distribution. *Mathematical Biosciences* 162:1–32.

Zheng, Q. 2005. New algorithms for Luria-Delbruck fluctuation analysis. *Mathematical Biosciences* 196:198–214.

Author Index

Subject Index